T0212474

# Lecture Notes in Business Information Processing 274

Series Editors

Wil M.P. van der Aalst
*Eindhoven Technical University, Eindhoven, The Netherlands*
John Mylopoulos
*University of Trento, Trento, Italy*
Michael Rosemann
*Queensland University of Technology, Brisbane, QLD, Australia*
Michael J. Shaw
*University of Illinois, Urbana-Champaign, IL, USA*
Clemens Szyperski
*Microsoft Research, Redmond, WA, USA*

More information about this series at http://www.springer.com/series/7911

Deepinder Bajwa · Sabine T. Koeszegi
Rudolf Vetschera (Eds.)

# Group Decision and Negotiation

## Theory, Empirical Evidence, and Application

16th International Conference, GDN 2016
Bellingham, WA, USA, June 20–24, 2016
Revised Selected Papers

 Springer

*Editors*
Deepinder Bajwa
Western Washington University
Bellingham, WA
USA

Rudolf Vetschera
University of Vienna
Vienna
Austria

Sabine T. Koeszegi ⓘ
TU Wien
Vienna
Austria

ISSN 1865-1348          ISSN 1865-1356   (electronic)
Lecture Notes in Business Information Processing
ISBN 978-3-319-52623-2          ISBN 978-3-319-52624-9   (eBook)
DOI 10.1007/978-3-319-52624-9

Library of Congress Control Number: 2016963652

© Springer International Publishing AG 2017
This work is subject to copyright. All rights are reserved by the Publisher, whether the whole or part of the material is concerned, specifically the rights of translation, reprinting, reuse of illustrations, recitation, broadcasting, reproduction on microfilms or in any other physical way, and transmission or information storage and retrieval, electronic adaptation, computer software, or by similar or dissimilar methodology now known or hereafter developed.
The use of general descriptive names, registered names, trademarks, service marks, etc. in this publication does not imply, even in the absence of a specific statement, that such names are exempt from the relevant protective laws and regulations and therefore free for general use.
The publisher, the authors and the editors are safe to assume that the advice and information in this book are believed to be true and accurate at the date of publication. Neither the publisher nor the authors or the editors give a warranty, express or implied, with respect to the material contained herein or for any errors or omissions that may have been made. The publisher remains neutral with regard to jurisdictional claims in published maps and institutional affiliations.

Printed on acid-free paper

This Springer imprint is published by Springer Nature
The registered company is Springer International Publishing AG
The registered company address is: Gewerbestrasse 11, 6330 Cham, Switzerland

# Preface

Group decision and negotiation (GDN) is a very broad field of research that deals with the many facets of decision-making processes involving multiple parties, from teams who want to combine their knowledge and ideas to find the solution that best satisfies their common goals, to adversaries trying to solve long-standing fundamental conflicts. Consequently, this field has attracted researchers from many disciplines such as operations research, economics, and political science, but also social sciences and communication, as well as information systems. Bringing researchers from these different disciplines together to share their insights and ideas on their common subject, discover similarities and complementarities of their research methodologies, and contribute to the common goal of a better understanding and support of these processes is in itself a difficult and important act of group decision-making and sometimes negotiation.

An important focal point for the GDN research community is the series of annual Group Decision and Negotiation conferences that started in 2000 with the Group Decision and Negotiation conference held in Glasgow, Scotland. Although it was originally planned as a one-time event, this conference was followed by a series of conferences that were held every year (with a single exception of 2011, when a planned conference in Jordan had to be canceled because of the turbulent situation in the region), and on four continents. Until 2015, ten conferences were held in Europe (Glasgow 2000, La Rochelle 2001, Istanbul 2003, Vienna 2005, Karlsruhe 2006, Coimbra 2008, Delft 2010, Stockholm 2013, Toulouse 2014, and Warsaw 2015), three in North America (Banff 2004, Mt. Tremblant 2007, and Toronto 2009), and one each in Australia (Perth 2002) and in South America (Recife 2012). Some of these meetings were held as streams within a larger INFORMS conference, but most were organized as separate events.

In 2016, GDN returned again to North America, and for the first time came to the United States. The Group Decision and Negotiation 2016 conference was hosted by Western Washington University in Bellingham, WA, and took place during June 20–24, 2016. In total, 70 papers were accepted for presentation at the conference after a first review process by members of the Program Committee (PC) and additional experts invited by the PC members. Following the tradition established in Toulouse 2014, two volumes of proceedings were created from the conference papers. Based on the results of the first of reviews, out of the 70 papers accepted for the conference, 12 papers were selected for publication in this volume of the *Lecture Notes in Business Information Processing*. After the conference, authors had the opportunity to revise their papers to take into account comments from the first round of reviews and to incorporate topics that might have arisen in discussions during the conference. These revised papers were then sent out for a second round of reviews, and in some cases this led to yet another revision of the papers. Papers not included in this volume were published in the electronic conference proceedings by Western Washington University.

We have grouped the papers included in this volume into four broad areas. The first group of papers mainly deals with the fundamental part of all decision processes, individual preferences. In group decisions and negotiations, like in many other areas of decision-making, preferences are often not clear at the outset when solving a decision problem, and this problem is aggravated when parties in a group decision or negotiation context do not only act on their behalf, but represent some organization or constituency. In such a setting, preferences of the organization or constituency need to be communicated to negotiators. In the first section of this volume, the first paper deals with these problems. In their paper "The Application of Item Response Theory for Analyzing the Negotiators' Accuracy in Defining Their Preferences," Ewa Roszkowska and Tomasz Wachowicz present empirical evidence of how difficult it might be to communicate an organization's preferences clearly to negotiators acting on behalf of the organization, and introduce item response theory as an instrument that might help organizations to identify issues and forms of communication that are particularly prone to misunderstandings. The second paper in this section, "Trade-Offs for Ordinal Ranking Methods in Multi-criteria Decisions" by Mats Danielson and Love Ekenberg deals with the problem that parties in a group decision context might not be able to specify the importance of criteria to be negotiated exactly, and surrogate methods have to be used to quantify the attribute weights.

The second section of this volume contains papers related to situations of group decision-making in which there is not necessarily a strong conflict of interests between group members, but different expertise and information as well as some different perspectives of the problem need to be integrated. Pascale Zaraté, Guy Camilleri, and D. Marc Kilgour in their paper "Multi-criteria Group Decision-Making with Private and Shared Criteria: An Experiment" deal with this mixture of common and individual interests. If a decision problem involves multiple criteria, some of these criteria might be seen in exactly the same way by all group members, but they might have different views on other criteria. This paper provides the first empirical evidence on how such shared criteria might influence the process of group decision-making. Any form of group decision-making requires some rules on how to aggregate different preferences and opinions. The literature offers many possible rules, and this obviously makes the choice of a rule to be used by the group another (meta) group decision problem. In their paper "Plurality, Borda Count, or Anti-plurality: Regress Convergence Phenomenon in the Procedural Choice," Takahiro Suzuki and Masahide Horita show that this does not necessarily lead to an infinite regress (decide about the rule to choose a rule to decide about a rule to choose a rule...), but that this process can converge at some level. We also want to mention that for this paper, the first author Takahiro Suzuki won the best young researcher's award at the conference.

The following two papers in this section focus on processes of group decision-making and on empirical methods to study these processes. Often, group decisions are made under time pressure, and then heuristic approaches are applied in decision-making. In their paper "Estimating Computational Models of Dynamic Decision-Making from Transactional Data," James Brooks, David Mendonça, Xin Zhang, and Martha Grabowski present a method of how parameters of such processes can be inferred from decisions observed in a highly volatile environment. Log data tracing a decision process over time also forms the empirical basis of the paper

"Demystifying Facilitation: A New Approach to Investigating the Role of Facilitation in Group Decision Support Processes" by Mike Yearworth and Leroy White. They describe how data recorded by a web-based group decision support system can be used to study how facilitators can actually influence the group process in such an environment, where facilitators are not physically present and share electronic communication channels in exactly the same way as all other group members.

The next section presents papers that study collective decision-making in situations characterized by a higher level of conflict, in particular negotiations. All three papers in this section deal with the measurement of important concepts in negotiations. In their paper "Bargaining Power: Measuring Its Drivers and Consequences in Negotiations," Tilman Eichstädt, Ali Hotait, and Niklas Dahlen are concerned with the issue of power. They relate standard concepts of negotiation analysis like the BATNA (best alternative to negotiated agreement) that a negotiator has to their power in the negotiation and finally to the outcome these negotiators achieve, and analyze these relationships in a controlled experiment. Negotiators do not always act rationally, and thus do not always reach the theoretically optimal outcomes. How to measure this deviation is the topic of the paper "A Deviation Index Proposal to Evaluate Group Decision-Making Based on Equilibrium Solutions" by Alexandre Bevilacqua Leoneti and Fernanda de Sessa. Emotions are also an important factor influencing the negotiation process, but existing methods to measure emotions in negotiations require considerable effort by raters. Michael Filzmoser, Sabine T. Koeszegi, and Guenther Pfeffer study whether methods of automatic text mining can be used for this task in their paper "What Computers Can Tell Us About Emotions: Classification of Affective Communication in Electronic Negotiations by Supervised Machine Learning."

The last section of this volume contains three papers related to group processes and negotiations in different subject areas. The first paper in this section "Facebook and the Elderly: The Benefits of Social Media Adoption for Aged Care Facility Residents" by Saara Matilainen, David G. Schwartz, and John Zeleznikow deals with an important aspect of group processes, connectedness, in the context of social networks and their use by the elderly population. An innovative approach to group decision-making is presented in the paper "How to Help a Pedagogical Team of an MOOC Identify the 'Leader Learners'?" by Sarra Bouzayane and Inès Saad, who describe a group decision support tool based on rough set theory and its application in a specific problem of online education. Last, but not least, the paper "Negotiating Peace: The Role of Procedural and Distributive Justice in Achieving Durable Peace" by Daniel Druckman and Lynn Wagner considers important aspects of negotiations in a political and diplomatic context. This paper won the best paper award at the conference.

Of course, organizing a conference like GDN and preparing such a volume of proceedings is not possible without many helping hands. Like everyone in the GDN community, we are deeply indebted to Mel Shakun, the founder of both the GDN section and the journal, for his continuing support, advice, and inspiration that has shaped our community for so many years. We also want to thank the two general chairs of GDN 2016, Colin Eden and Gregory Kersten, for their many contributions to the success of the conference, and the local organizers, in particular Marlene Harlan and her team, whose effort made GDN 2016 possible.

Papers in this volume went through an elaborate two-stage review process, and the quality and timeliness of reviews were essential for the preparation of this volume. We therefore are very grateful to all reviewers of the papers. Our thanks goes to: Fran Ackerman, Adiel Almeida, Adiel Almeida Filho, Jonatas Almeida, Ana-Paula Cabral, Colin Eden, Alberto Franco, Johannes Gettinger, Salvatore Greco, Masahide Horita, Gregory Kersten, Hsiangchu Lai, Annika Lenz, Bilyana Martinovski, Danielle Morais, José Maria Moreno-Jiménez, Hannu Nurmi, Amer Obeidi, Mareike Schoop, Ofir Turel, Doug Vogel, Tomasz Wachowicz, Shi Kui Wu, Bo Yu, Yufei Yuan, Pascale Zaraté, and John Zeleznikow.

Finally, we thank Ralf Gerstner, Alfred Hofmann, and Christine Reiss at Springer publishers for the excellent collaboration, Iurii Berlach and Ruth Strobl for their technical assistance, and, last but definitely not the least, our families who (in particular on the Vienna side of this collaboration) had to endure many late-night Skype meetings in preparing GDN 2016.

December 2016

Deepinder Bajwa
Sabine Koeszegi
Rudolf Vetschera

# Organization

## Honorary Chair

Melvin Shakun          New York University, USA

## General Chairs

Colin Eden          Strathclyde University, Glasgow, UK
Gregory Kersten          Concordia University, Montreal, Canada

## Program Chairs

Deepinder Bajwa          Western Washington University, USA
Sabine Koeszegi          Vienna University of Technology, Austria
Rudolf Vetschera          University of Vienna, Austria

## Organizing Chair

Marlene Harlan          Western Washington University, USA

## Program Committee

| | |
|---|---|
| Fran Ackermann | Curtin University, Australia |
| Adiel Almeida | Federal University of Pernambuco, Brazil |
| Martin Bichler | Technical University of Munich, Germany |
| Xusen Cheng | University of International Business, China |
| João C. Clímaco | University of Coimbra, Portugal |
| Suzana F.D. Daher | Federal University of Pernambuco, Brazil |
| Gert-Jan de Vrede | University of Nebraska-Omaha, USA |
| Luis C. Dias | University of Coimbra, Portugal |
| Liping Fang | Ryerson University, Canada |
| Raimo Pertti Hämäläinen | Aalto University, Finland |
| Keith Hipel | University of Waterloo, Canada |
| Masahide Horita | University of Tokyo, Japan |
| Takayuki Ito | Nagoya Institute of Technology, Japan |
| Catholjin Jonker | Delft University of Technology, The Netherlands |
| Marc Kilgour | Wilfrid Laurier University, Canada |
| Kevin Li | University of Windsor, Canada |
| Bilyana Martinovski | Stockholm University, Sweden |
| Paul Meerts | Clingendael Institute, The Netherlands |
| Ugo Merlone | University of Turin, Italy |
| Danielle Morais | Federal University of Pernambuco, Brazil |

| José Maria Moreno-Jiménez | Zaragoza University, Spain |
| Hannu Nurmi | University of Turku, Finland |
| Amer Obeidi | University of Waterloo, Canada |
| Pierpaolo Pontradolfo | Politecnico di Bari, Italy |
| Ewa Roszkowska | University of Białystok, Poland |
| Mareike Schoop | Hohenheim University, Germany |
| Wei Shang | Chinese Academy of Sciences, China |
| Rangaraja Sundraraj | Indian Institute of Management, India |
| Katia Sycara | Carnegie Mellon University, USA |
| Tomasz Szapiro | Warsaw School of Economics, Poland |
| Przemyslaw Szufel | Warsaw School of Economics, Poland |
| David P. Tegarden | Virginia Tech, USA |
| Ernest M. Thiessen | SmartSettle, Canada |
| Ofir Turel | California State University, USA |
| Rustam Vahidov | Concordia University, Canada |
| Doug Vogel | Harbin Institute of Technology, China |
| Tomasz Wachowicz | University of Economics Katowice, Poland |
| Christof Weinhardt | Karlsruhe Institute of Technology, Germany |
| Shi Kui Wu | University of Windsor, Canada |
| Yufei Yuan | McMaster University, Canada |
| Bo Yu | Concordia University, Canada |
| Pascale Zaraté | University Toulouse 1, France |
| John Zeleznikow | Victoria University, Australia |

## Organizing Committee

| Michele Anderson | Western Washington University, USA |
| Logan Green | Western Washington University, USA |
| Courtney Hiatt | Western Washington University, USA |
| Mina Tanaka | Western Washington University, USA |
| Craig Tyran | Western Washington University, USA |
| Xiofeng Chen | Western Washington University, USA |
| Joseph Garcia | Western Washington University, USA |
| Jean Webster | Western Washington University, USA |

# Contents

## Applications of Group Decision and Negotiation

# Individual Preferences in Group Decision and Negotiation

# The Application of Item Response Theory for Analyzing the Negotiators' Accuracy in Defining Their Preferences

Ewa Roszkowska[1]([⊠]) [iD] and Tomasz Wachowicz[2] [iD]

[1] Faculty of Economy and Management,
University of Bialystok, Warszawska 63, 15-062 Bialystok, Poland
erosz@o2.pl
[2] Department of Operations Research, University of Economics in Katowice,
1 Maja 50, 40-287 Katowice, Poland
tomasz.wachowicz@ue.katowice.pl

**Abstract.** In this paper we analyze how some notions of Item Response Theory (IRT) may be used to analyze the process of scoring the negotiation template and building the negotiation offer scoring system. In particular we focus on evaluating and analyzing the accuracy and concordance of such scoring systems with the preferential information provided to negotiators by the represented party. In our research we use the dataset of bilateral electronic negotiations conducted by means of Inspire negotiation support system, which provides users with decision support tools for preference analysis and scoring system building based on SMART/SAW method. IRT allows us to consider how the potential accuracy of individual scoring systems can be explained by both negotiators' intrinsic abilities to use decision support tool and understand the scoring mechanism, and the difficulty of applying this scoring mechanism.

**Keywords:** Preference elicitation and analysis · Negotiation offer scoring systems · Ratings accuracy · Item Response Theory · Rasch Model

## 1 Introduction

From the viewpoint of decision support in multi-issue negotiations the negotiation offer scoring systems play a key role. Negotiation offer scoring system is a formal rating system that takes into account the negotiator's individual preferences allowing to evaluate any feasible negotiation offer that fits the negotiation template [1]. It is used in the actual negotiation phase to support individual negotiators as it compares the offers submitted to the negotiation table, measures the scale of concessions made by the parties, depicts the negotiation history graphs or generates pro-active alternative negotiation solutions (see [2, 3]). When determined for both parties, scoring systems may also be used to perform a symmetric negotiation analysis, i.e. to verify efficiency of the negotiation agreement and analyze its fairness in comparison to selected notions of the bargaining solution [4, 5]. Providing negotiators with additional information about offers' quality and scales of concession may significantly influence the negotiators' perception of the flow of the negotiation process and the interpretation of the

© Springer International Publishing AG 2017
D. Bajwa et al. (Eds.): GDN 2016, LNBIP 274, pp. 3–15, 2017.
DOI: 10.1007/978-3-319-52624-9_1

counterparts' moves. It is then crucial to ensure that such scoring system be built individually by the negotiator to reflect accurately her preferences; this makes it a reliable tool for decision support in a negotiation.

There are, however, various factors that may determine negotiators' accuracy in building their scoring systems. One of them is the decision and cognitive abilities (limitations) of negotiators [6], which – in our case – determine their general skills in performing the formal pre-negotiation analysis required to elicit their preferences and build a negotiation offer scoring system. Second is the complexity of the decision support tool that is offered to the negotiators to facilitate such a process of pre-negotiation analysis. Various multiple criteria decision making (MCDM) methods may be used [7–9] to support negotiators in building the negotiation offer scoring system. However, the most frequently used ones derive mainly from the idea of additive weighting [10, 11], which, in the case of discrete decision making and negotiation problems, amounts to a simple assignment of rating points to the elements of the negotiation template, using a well know and straightforward SMART algorithm [12] (*Simple Multi-Attribute Rating Technique*). The latter, which is claimed to be technically the simplest decision support tool, is applied in such negotiation support systems as Inspire [2], SmartSettle [13] or, in a simplified form and among other preference elicitation techniques, in Negoisst [3]. Thus, it seems theoretically interesting and pragmatically vital to verify how effectively the simplest SMART-based scoring systems are used by negotiators to provide an accurate pre-negotiation analysis and build reliable scoring systems used in decision support in subsequent negotiation phases.

Earlier research focused on analyzing the use and usefulness of scoring systems, and on measuring the applicability of SMART- or SAW-based tools for supporting negotiators [14–16]. This paper is a part of bigger experimental research that studies the accuracy of SMART-based scoring systems, as well as the determinants and consequences of such an accuracy [17, 18]. The goal of this paper is to analyze the links between the negotiators' abilities to accurately map the preferential information into a system of cardinal ratings and the difficulty of the decision support mechanism designed to facilitate them in this task. To study the potential relationships we use the dataset of the electronic bilateral negotiations conducted by means of Inspire negotiation support system [2] and analyze it using the parameter model of Item Response Theory (IRT) [19]. We treat a SMART-based rating procedure as a series of questions required by IRT for which the answers are ratings the negotiators have assigned to issues and options. The accuracy of answers is verified by evaluating the concordance between the ratings assigned by the negotiator to options and issues and the ranking of these options and issues described in the negotiation case (which we call the ordinal accuracy of the scoring system). This allows us to calculate the ordinal accuracy indexes of negotiators' individual ratings and analyze how they are influenced by negotiators' abilities and the complexity of the assignment procedure.

The paper consists of four more sections. In Sect. 2 we describe the fundamentals of IRT and discuss the Rasch Model that we applied in our research. In Sect. 3 we describe the Inspire negotiation support system and its procedure for building the negotiation offers scoring system. We also define the notion of ordinal accuracy of such scoring systems that will be applied in our IRT model. In Sect. 4 we discuss the results. Section 5 suggests future work.

## 2 Fundamentals of Item Response Theory

Item Response Theory (IRT) is a general framework for specifying mathematical functions that describe the interactions and the relationship between persons taking part in tests (surveys) and the test items [19]. IRT assumes that some traits of a person are latent, i.e. they cannot be measured directly. However, they influence the measurable and observable performance of the examinees. There may be also another factor that affects an individual's probability of responding to an item in a particular way, namely the level of difficulty of this item. An individual who has a high level of ability will be more likely to respond correctly to an item than an individual who has a low level of ability. Also, a difficult item requires a relatively high trait level so that it may be answered correctly, but an easy item requires only a low trait level to be answered correctly. Although they are separate issues, in the IRT analysis the trait level and the item difficulty are intrinsically connected, so the item difficulty is expressed in terms of trait level.

There is a number of different IRT models [19], among which one-parameter logistic Rasch Model [20] – predicts the probability of a correct response to a binary item (i.e., right/wrong, true/false, agree/disagree) based on interaction between the individual ability $\theta$ and the item parameter $b$ (difficulty parameter). Both the trait levels and the item difficulties are usually scored on a standardized metric, so that their means are 0 and the standard deviations are 1. The main elements of Rasch Model are the following:

1. The *item response function* is a probability of correct response given to the $i$ th item by an examinee with an individual ability level $\theta$. It is presented in a following form:

$$P_i(\theta, b_i) = \frac{\exp(\theta - b_i)}{1 + \exp(\theta - b_i)}, \tag{1}$$

where: $\theta -$ the individual ability level, $b_i$ – item difficulty parameter (constant).

2. Any item in a test provides some information about the examinee's ability, but the amount of this information depends on how closely the difficulty of the item matches the ability of the person. The *item information function* $I_i(\theta, b_i)$ of the Rasch Model is the following [21]:

$$I_i(\theta, b_i) = P_i(\theta, b_i) \cdot Q_i(\theta, b_i) = P_i(\theta, b_i) \cdot (1 - P_i(\theta, b_i)), \tag{2}$$

where: $Q_i(\theta, b_i) -$ is the probability of wrong response to $i$ th item given by examinee whose ability is $\theta$.

3. Any person who takes a test of $k$ items can have one of $k+1$ possible scores by getting 0, 1, 2, 3, 4, ..., $k$ items right. The *test response function* for examinees with ability $\theta$ is the sum of the item response functions for all items in the test:

$$P(\theta) = \sum_i P_i(\theta, b_i). \tag{3}$$

Similarly the *test information function* can be expressed as:

$$I(\theta) = \sum_i I_i(\theta, b_i). \tag{4}$$

The *standard error of measurement* (SEM) reflects the accuracy to which Rasch Model can measure any value of the latent ability and has the following form:

$$SEM(\theta) = \sqrt{\frac{1}{I(\theta)}}. \tag{5}$$

## 3 SMART-Based Negotiation Offer Scoring System and Its Accuracy

In our research we implement the Rasch Model to analyze the negotiators' ability to build accurate negotiation offer scoring systems in electronic negotiations. To conduct the experiments we used the Inspire negotiation support system [2]. Instead of analyzing classic tests with a series of questions/items (as assumed in IRT), we study the process of building the scoring system by means of SMART-based procedure implemented in Inspire and measure the correctness of negotiators at each stage of such scoring procedure. Therefore, the negotiation template and the predefined preferential information need to be specified for an experiment, which we treat as the references in verifying the correctness of the scoring system individually built by the negotiators. One of Insire's predefined cases that fulfills the above requirements is the Mosico-Fado case, in which a contract between an entertainment company – Mosico, and the singer – Fado is negotiated. In this case the negotiation template is defined by the means of four issues, each having a predefined list of salient options (see Table 1).

**Table 1.** Mosico-Fado negotiation template.

| Issues to negotiate | Issue options |
|---|---|
| Number of new songs (introduced and performed each year) | 11; 12; 13; 14 or 15 songs |
| Royalties for CDs (in percent) | 1.5; 2; 2.5 or 3% |
| Contract signing bonus (in dollars) | $125,000; $150,000; $200,000 |
| Number of promotional concerts (per year) | 5; 6; 7 or 8 concerts |

In Mosico-Fado case the structures of preferences for both parties are described both verbally and graphically. An example of preference description for Fado party is presented in Fig. 1.

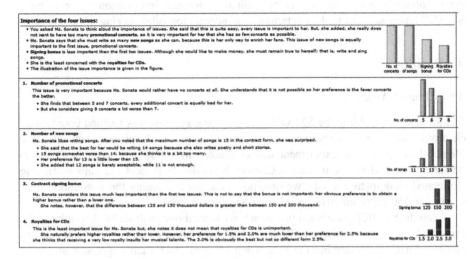

**Fig. 1.** Verbal and graphical representation of preferences in Inspire

The concordance of negotiators' individual offer scoring systems with the preferential information presented above may be measured by means of two notions of ordinal and cardinal accuracy (see [18]). For the purpose of Rasch Model the former is used; it checks if the negotiators follow the order of preferences depicted by the bar sizes both for the issue and option ratings. Namely, if $m$ issues (or options) $A_1, \ldots, A_m$ are ordered according to decreasing preferences required by ordinal accuracy the corresponding ratings $u(A_i)$ assigned to these issues should satisfy the following condition

$$u(A_1) > u(A_2) > \ldots > u(A_m). \tag{6}$$

The ordinal accuracy of the scoring system built by $j$th negotiator is represented by a ratio of the number of correct ratings ($n_j^{\text{cor}}$), i.e., the ratings that satisfy the condition (6), and the total number ($n$) of all the rankings to be built within the negotiation template:

$$OA_j = \frac{n_j^{\text{cor}}}{n}. \tag{7}$$

In the problem described in Fig. 1 we have $n = 5$ because one ranking represents the issue importance and four other rankings represent the orders of options within each issue respectively. Note, that from the viewpoint of IRT analysis $n$ represents the number of items (questions) that need to be answered by the negotiators. If the ratings assigned by the negotiator to the evaluated elements of the template (i.e. set of issues or four sets of options) reflect the ranking (bars' height) correctly, the item is considered to be answered correctly (1), otherwise we considered this item to have wrong answer (0). Consequently, a latent trait, i.e. an underlying unobservable variable, is defined as the negotiator's ability of preserving ordinal preference information while using a SMART-based support tool to determine the negotiation offer scoring system.

## 4  Results

### 4.1  Dataset Analysis

The accuracy in building the SMART-based scoring systems was verified based on an electronic negotiation experiment conducted using the Inspire system and the Mosico-Fado case. The experiment was conducted in spring 2015 and the 332 participants were students from Austria, Canada, Netherlands, Poland and Taiwan. Having eliminated incomplete records, we obtained a dataset containing the negotiation transcripts for 150 representatives of Mosico and 161 of Fado.

According to IRT estimation procedure, we started our analysis by determining the trait level for each negotiator [22]. The initial estimation of trait levels is a two-step process. First, we determine the proportion of items that each negotiator answered correctly. Note that it is defined exactly by the ordinal accuracy index $OA_j$ defined by formula (7). Next, the logarithm of the ratio of correct and incorrect answers needs to be determined:

$$\theta_j = \ln\left(\frac{OA_j}{1 - OA_j}\right). \tag{8}$$

Similarly, we can estimate item difficulties. First, we determine the proportion of correct responses $P_i$ for each item $i$, and then compute the logarithm of the ratio of the proportion of incorrect responses to the proportion of correct responses:

$$b_i = \ln\left(\frac{1 - P_i}{P_i}\right). \tag{9}$$

The values of initial estimations of $\theta_j$ and $b_i$ determined for various classes of negotiators' ordinal accuracy are presented in Table 2 (Fados) and Table 3 (Mosicos).

Since the item difficulty parameters and personal trait levels usually do not fit the item response functions given by formula (1), they need to be adjusted by means of maximum likelihood estimation method (CML) [22]. Thus, in the next step, the initial data from Tables 2 and 3 is processed to estimate the final values of parameters of the item response function. The result of this iterative procedure is presented in Table 4.

Figure 2 presents the item response functions of five items for Fado and Mosico conforming to the Rasch Model and determined according to formula (1).

Observe that an item's difficulty depends of the role in negotiation. For Fados the most difficult was to assign correct ratings to weights (Item 1), whereas for Mosicos –

**Table 2.** Initial estimates of item difficulty and Fados' trait levels.

| Fado class (no. of neg.) | Answers | | | | | Proportion $(OA_i)$ | Trait level $(\theta_j)$ |
| --- | --- | --- | --- | --- | --- | --- | --- |
| | Item 1 | Item 2 | Item 3 | Item 4 | Item 5 | | |
| 1 (N = 12) | 0 | 0 | 0 | 0 | 0 | 0.0 | |
| 2 (N = 6) | 0 | 1 | 0 | 0 | 0 | 0.2 | −1.386 |
| 3 (N = 2) | 0 | 0 | 1 | 0 | 0 | 0.2 | −1.386 |
| 4 (N = 3) | 0 | 0 | 0 | 1 | 0 | 0.2 | −1.386 |
| 5 (N = 4) | 0 | 0 | 0 | 0 | 1 | 0.2 | −1.386 |
| 6 (N = 1) | 1 | 0 | 0 | 0 | 0 | 0.2 | −1.386 |
| 7 (N = 3) | 0 | 1 | 1 | 0 | 0 | 0.4 | −0.405 |
| 8 (N = 5) | 0 | 1 | 0 | 0 | 1 | 0.4 | −0.405 |
| 9 (N = 5) | 0 | 0 | 0 | 1 | 1 | 0.4 | −0.405 |
| 10 (N = 1) | 1 | 1 | 0 | 0 | 0 | 0.4 | −0.405 |
| 11 (N = 1) | 0 | 1 | 0 | 1 | 0 | 0.4 | −0.405 |
| 12 (N = 2) | 1 | 0 | 0 | 1 | 1 | 0.6 | 0.405 |
| 13 (N = 20) | 0 | 1 | 0 | 1 | 1 | 0.6 | 0.405 |
| 14 (N = 1) | 0 | 1 | 1 | 0 | 1 | 0.6 | 0.405 |
| 15 (N = 6) | 0 | 0 | 1 | 1 | 1 | 0.6 | 0.405 |
| 16 (N = 2) | 1 | 1 | 0 | 1 | 1 | 0.8 | 1.386 |
| 17 (N = 2) | 1 | 0 | 1 | 1 | 1 | 0.8 | 1.386 |
| 18 (N = 1) | 1 | 1 | 1 | 0 | 1 | 0.8 | 1.386 |
| 19 (N = 48) | 0 | 1 | 1 | 1 | 1 | 0.8 | 1.386 |
| 20 (N = 36) | 1 | 1 | 1 | 1 | 1 | 1.0 | |
| Prop. correct | 0.28 | 0.77 | 0.62 | 0.78 | 0.82 | | |
| Difficulty $b_i$ | 0.94 | −1.21 | −0.49 | −1.27 | −1.52 | | |

royalties options (Item 4). For Fados the easiest ratings to assign were bonus options (Item 5), whereas for Mosicos – the number of concerts (Item 2). Higher difficulty levels indicate that higher trait (ability) levels are required for negotiator to have a 50% chance to "answer correctly". For instance, a Fado representative with a trait level of 2.80 (i.e. the ability of mapping ordinal preferences accurately into a system of cardinal scores is higher than average of 2.80 standard deviation value) will have a 50% chance of rating the issue weights (Item 1) correctly because Item 1's difficulty level is equal to 2.80.

Looking at the curves, we can see that Fado with an average level of ability of preserving ordinal preference information has about 0.057 chance to rate correctly the weights, 0.709 chance rate correctly the number of concerts option, 0.381 chance to rate correctly the number of songs option, 0.722 chance to rate correctly royalties and 0.809 chance to rate correctly the signing bonus. For the Mosico the probabilities are: 0.588 for weights, 0.778 for concerts, 0.25 for songs, 0.182 for royalties and 0.731 for contract.

Now we use IRT to estimate the psychometric quality of the test across a wide range of trait levels. This can be seen as a two-step process again. First, we evaluate the psychometric quality of each item across a range of trait levels, i.e. $I_i(\theta, b_i)$. Figure 3

**Table 3.** Initial estimates of item difficulty and Mosicos' trait levels.

| Fado class (no. of neg.) | Answers | | | | | Proportion $(OA_i)$ | Trait level $(\theta_j)$ |
|---|---|---|---|---|---|---|---|
| | Item 1 | Item 2 | Item 3 | Item 4 | Item 5 | | |
| 1 (N = 6) | 0 | 0 | 0 | 0 | 0 | 0.0 | |
| 2 (N = 10) | 0 | 1 | 0 | 0 | 0 | 0.2 | −1.386 |
| 3 (N = 1) | 0 | 0 | 1 | 0 | 0 | 0.2 | −1.386 |
| 4 (N = 3) | 0 | 0 | 0 | 0 | 1 | 0.2 | −1.386 |
| 5 (N = 4) | 1 | 0 | 0 | 0 | 0 | 0.2 | −1.386 |
| 6 (N = 2) | 1 | 0 | 0 | 0 | 1 | 0.4 | −0.405 |
| 7 (N = 2) | 0 | 1 | 1 | 0 | 0 | 0.4 | −0.405 |
| 8 (N = 9) | 0 | 1 | 0 | 0 | 1 | 0.4 | −0.405 |
| 9 (N = 1) | 0 | 0 | 0 | 1 | 1 | 0.4 | −0.405 |
| 10 (N = 4) | 1 | 1 | 0 | 0 | 0 | 0.4 | −0.405 |
| 11 (N = 18) | 1 | 1 | 0 | 0 | 1 | 0.6 | 0.405 |
| 12 (N = 6) | 1 | 0 | 0 | 1 | 1 | 0.6 | 0.405 |
| 13 (N = 1) | 1 | 1 | 1 | 0 | 0 | 0.6 | 0.405 |
| 14 (N = 2) | 1 | 0 | 1 | 0 | 1 | 0.6 | 0.405 |
| 15 (N = 1) | 0 | 1 | 1 | 0 | 1 | 0.6 | 0.405 |
| 16 (N = 1) | 1 | 1 | 1 | 1 | 0 | 0.8 | 1.386 |
| 17 (N = 6) | 1 | 1 | 0 | 1 | 1 | 0.8 | 1.386 |
| 18 (N = 14) | 1 | 1 | 1 | 0 | 1 | 0.8 | 1.386 |
| 19 (N = 7) | 0 | 1 | 1 | 1 | 1 | 0.8 | 1.386 |
| 20 (N = 52) | 1 | 1 | 1 | 1 | 1 | 1.0 | |
| Prop. correct | 0.73 | 0.83 | 0.54 | 0.49 | 0.81 | | |
| Difficulty $b_i$ | −0.99 | −1.59 | −0.16 | 0.04 | −1.45 | | |

**Table 4.** The results of the revised estimation of item's difficulty parameters $\hat{b}_i$.

| Item | Item 1: Weights | | Item 2: Concerts | | Item 3: Songs | | Item 4: Royalties | | Item 5: Signing | |
|---|---|---|---|---|---|---|---|---|---|---|
| | Fado | Mosico | Fado | Mosico | Fado | Mosico | Fado | Mosico | Fado | Mosico |
| No. of correct answers | 45 | 110 | 124 | 125 | 99 | 81 | 125 | 73 | 132 | 121 |
| % of correct answers | 28.0 | 73.3 | 77.0 | 83.3 | 61.5 | 54.0 | 77.6 | 48.7 | 82.0 | 80.7 |
| Difficulty parameter $\hat{b}_i$ | 2.80 | −0.35 | −0.89 | −1.25 | 0.48 | 1.10 | −0.95 | 1.50 | −1.44 | −0.99 |
| S.E. | 0.26 | 0.25 | 0.26 | 0.27 | 0.23 | 0.24 | 0.26 | 0.25 | 0.28 | 0.26 |

Source: GANZ RASCH software results

presents the item information functions conforming to the Rasch Model for Fado and Mosico and determined based on formula (2).

Higher item information values indicate greater psychometric quality, i.e. the greater ability to differentiate the negotiators of a particular trait level. For example,

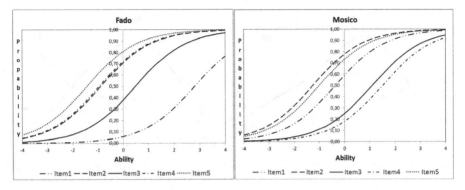

**Fig. 2.** Item response functions for Mosico and Fado conforming to the Rasch Model

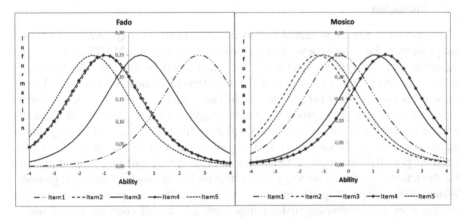

**Fig. 3.** Item information functions for Fado and Mosico conforming to the Rasch Model

items 2, 4, 5 for Fado and items 1, 2, 5 for Mosico have higher psychometric quality at relatively low trait levels. Thus, these items have greater capacity to discriminate among people with low trait levels than among people with high trait levels.

Now we can combine item information values to obtain test information values, i.e. $I(\theta)$. A test information curve is useful for illustrating the degree to which a test provides different quality of information at different trait levels. The test information function and standard error of measurement (SEM) determined based on formulas (4) and (5), are presented in Fig. 4.

Let us observe that the SEM function for Fado and Mosico is quite flat for abilities within the range $(-2;2)$ and increases for both smaller and larger abilities. Thus, our test successfully differentiates people who have trait levels within two standard deviations above and below the mean. For Fado's preferential information test provides the greatest information at the trait level within one standard deviation below the mean $(\theta \approx -1)$. For Mosico – test provides the best information at average trait level $(\theta = 0)$; it provides less information at more extreme trait levels.

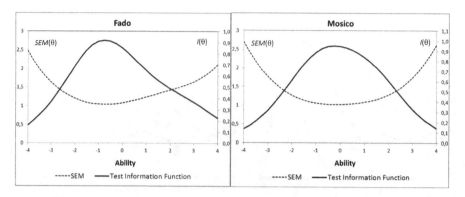

**Fig. 4.** Test information function and standard error of measurement for the Rasch Model

## 4.2   Discussion

The results we presented in Sect. 4.1 allow us to formulate a few conclusions regarding our problem of building accurate scoring systems by means of SMART-based scoring algorithm. The item response and item information functions show that the differences in the negotiators' abilities and item difficulties depend on the role (Mosico/Fado) the participants played in our experiment. For instance, analyzing Fig. 2 we see that most difficult for Fados was the process of rating weights (item 1), while for Mosicos the process of rating the resolution levels of royalties (item 4). Since issue rating differs technically from option rating (see first step of SMART procedure in [12]), we analyzed the potential reasons for different item 1 response functions for Fado and Mosico. We think that the potential difficulty of this item for Fados may have technical grounds. While Fados' graphical preferential information specifies the priorities over issues starting from most important to least important, i.e.: no. of concerts, no. of songs, *signing bonus and royalties* (see Fig. 1); the Inspire scoring form required assigning the ratings for issues listed in different order, i.e.: no. of concerts, no. of songs, *royalties and signing bonus*. As the data analysis showed, the vast majority of errors was made by Fado negotiators in assigning the higher ratings to royalties than to signing bonus. We contribute it to the direct effect of fast thinking that results from purely technical issue related to organizing the preferential information and scoring procedure, but not to the difficulty of the procedure as such. This conclusion is also confirmed by the position of the item information functions (Fig. 3). If the SMART-based scoring procedure had been considered an easy tool by the negotiators, all the item information functions would have reached their maximum for small ability values (at the left hand side of the graph). In our case, the function peaks are distributed over the whole range of abilities and vary for parties. It means that different abilities are required to perform similar rating tasks correctly. Thus, the difficulty of building a scoring is rather case-specific and depends on the nuances of structures of preferences. When interpreting the test information functions we found that our scoring procedure (a test of five questions) differentiates between good negotiators and negotiators with average abilities (peaks close to $\theta = 0$. It means that our "average users" have 50% chance to do ratings correctly. We do not know the source of such result (general low ability in the sample of negotiators or objectively high difficulty of scoring procedure) but having predefined preferential information and easy to use and user friendly scoring procedure allowing to determine accurate scoring systems, we would expect our test information function to reach its maximum for low abilities only – i.e. negotiators of low ability may do their ratings either correctly or not – depending on other factors to be identified by higher-parameters model; but those of average and high ability will on average always proceed correctly.

## 5  Summary

Using an IRT approach in studying the problem of building the negotiation offer scoring system allows us to analyze the probability of negotiators preserving the preferential information, while rating the elements of the negotiation template. What is interesting, it allows us to determine negotiators' abilities to do this task, yet not explicitly defined by means of a series of their characteristics but as a latent trait, which depends on both negotiator's responses to each task (correctly/incorrectly rated elements of the template) and the properties of the scoring process (difficulties of the rankings). What is important from the viewpoint of decision support in negotiation, the IRT test information function predicts the accuracy to which we can measure the latent ability. For practical reasons, knowledge of item difficulty can be useful for improvements in defining and presenting the individual preference information to assure the highest possible accuracy of the corresponding scoring systems.

In our study we used Rasch Model because it was the simplest IRT model that fits our research goal. However, there was also a technical reason for applying Rasch Model, namely the relatively small size of the dataset available. Various extensions of this basic model have been developed for more flexible modeling of different situations, such as: graded response models (GRM), which analyze ordinal responses and rating scales; two, three- and four-parameter models, which analyze test items that have guessing and carelessness parameters in the response curves; multidimensional IRT models, which analyze test items that can be explained by more than one latent trait or factor; or others [19]. Our future work will focus on implementing more advanced three- and four-parameter models that could allow us to identify if other factors, such as negotiators' carelessness (there are the students that have to participate, but they may not care about being diligent) play a role in the process of building the negotiation offer scoring systems. These models, however, require far bigger data samples, e.g. a three-parameter model requires at least 2000 observations to provide reliable results [19]. The future research will also require studying the potential impact of inaccurate scoring systems on negotiation process and outcomes.

**Acknowledgements.** This research was supported by the grant from Polish National Science Centre (2015/17/B/HS4/00941).

# References

1. Raiffa, H., Richardson, J., Metcalfe, D.: Negotiation Analysis: The Science and Art of Collaborative Decision Making. The Balknap Press of Harvard University Press, Cambridge (2002)
2. Kersten, G.E., Noronha, S.J.: WWW-based negotiation support: design, implementation, and use. Decis. Support Syst. 25(2), 135–154 (1999)
3. Schoop, M., Jertila, A., List, T.: Negoisst: a negotiation support system for electronic business-to-business negotiations in e-commerce. Data Knowl. Eng. 47(3), 371–401 (2003)
4. Raiffa, H.: Arbitration schemes for generalized two-person games. Ann. Math. Stud. 28, 361–387 (1953)
5. Nash, J.F.: The bargaining problem. Econometrica 18, 155–162 (1950)
6. Frederick, S.: Cognitive reflection and decision making. J. Econ. Perspect. 19, 25–42 (2005)
7. Salo, A., Hämäläinen, R.P.: Multicriteria decision analysis in group decision processes. In: Kilgour, D.M., Eden, C. (eds.) Handbook of Group Decision and Negotiation. Advances in Group Decision and Negotiation, vol. 4, pp. 269–283. Springer, Heidelberg (2010)
8. Wachowicz, T.: Decision support in software supported negotiations. J. Bus. Econ. 11(4), 576–597 (2010)
9. Mustajoki, J., Hämäläinen, R.P.: Web-HIPRE: global decision support by value tree and AHP analysis. INFOR. J. 38(3), 208–220 (2000)
10. Keeney, R.L., Raiffa, H.: Decisions with Multiple Objectives: Preferences and Value Trade-Offs. Wiley, New York (1976)
11. Churchman, C.W., Ackoff, R.L.: An approximate measure of value. J. Oper. Res. Soc. Am. 2(2), 172–187 (1954)
12. Edwards, W., Barron, F.H.: SMARTS and SMARTER: improved simple methods for multiattribute utility measurement. Organ. Behav. Hum. Dec. 60(3), 306–325 (1994)
13. Thiessen, E.M., Soberg, A.: SmartSettle described with the Montreal taxonomy. Group Decis. Negot. 12(2), 165–170 (2003)
14. Roszkowska, E., Wachowicz, T.: SAW-based rankings vs. intrinsic evaluations of the negotiation offers – an experimental study. In: Zaraté, P., Kersten, G.E., Hernández, J.E. (eds.) GDN 2014. LNBIP, vol. 180, pp. 176–183. Springer, Heidelberg (2014). doi:10.1007/978-3-319-07179-4_20
15. Vetschera, R.: Preference structures and negotiator behavior in electronic negotiations. Decis. Support Syst. 44(1), 135–146 (2007)
16. Roszkowska, E., Wachowicz, T.: Defining preferences and reference points – a multiple criteria decision making experiment. In: Zaraté, P., Kersten, G.E., Hernández, J.E. (eds.) GDN 2014. LNBIP, vol. 180, pp. 136–143. Springer, Heidelberg (2014). doi:10.1007/978-3-319-07179-4_15
17. Kersten, G.E., Roszkowska, E., Wachowicz, T.: Do the negotiators' profiles influence an accuracy in defining the negotiation offer scoring systems. In: The 15th International Conference on Group Decision and Negotiation Letters, pp. 129–138. Warsaw School of Economics Press (2015)
18. Roszkowska, E., Wachowicz, T.: Inaccuracy in defining preferences by the electronic negotiation system users. In: Kamiński, B., Kersten, G.E., Szapiro, T. (eds.) GDN 2015. LNBIP, vol. 218, pp. 131–143. Springer, Heidelberg (2015). doi:10.1007/978-3-319-19515-5_11
19. Hambleton, R.K., Swaminathan, H.: Item Response Theory: Principles and Applications. Springer Science & Business Media, Berlin (1985)

20. Rasch, G.: An item analysis which takes individual differences into account. Br. J. Math. Stat. Psychol. **19**(1), 49–57 (1966)
21. Embretsson, S., Reise, S.: Item Response Theory for Psychologists. Lawrence Erlbaum Associates, Publishers, Mahwah (2000)
22. Alexandrowicz, R.W.: "GANZ RASCH" a free software for categorical data analysis. Soc. Sci. Comput. Rev. **30**(3), 369–379 (2012)

# Trade-Offs for Ordinal Ranking Methods in Multi-criteria Decisions

Mats Danielson[1,2] and Love Ekenberg[1,2(✉)] (iD)

[1] Department of Computer and Systems Sciences,
Stockholm University, Box 7003, 164 07 Kista, Sweden
mats.danielson@su.se, ekenberg@iiasa.ac.at
[2] International Institute of Applied Systems Analysis, IIASA,
Schlossplatz 1, 2361 Laxenburg, Austria

**Abstract.** Weight elicitation methods in multi-criteria decision analysis (MCDA) are often cognitively demanding, require too much precision and too much time and effort. Some of the issues may be remedied by connecting elicitation methods to an inference engine facilitating a quick and easy method for decision-makers to use weaker input statements, yet being able to utilize these statements in a method for decision evaluation. One important class of such methods ranks the criteria and converts the resulting ranking into numerical so called surrogate weights. We analyse the relevance of these methods and discuss how robust they are as candidates for modelling decision-makers and analysing multi-criteria decision problems under the perspectives of several stakeholders.

**Keywords:** Multi-criteria decision analysis · Criteria weights · Criteria ranking · Rank order

## 1 Introduction

Regardless of the methods used for decision making, there exists a real problem in that numerically precise information is seldom available, and when it comes to providing reasonable weights in multi-criteria decision problems, there are severe difficulties due to decision-makers not seeming to have neither precise information at hand nor the required discrimination capacity. The same problem appears in group decision settings, where there is a desire to rank or in other ways compare the views or values of different participants or stakeholders and there are several approaches to this. For instance, Fig. 1 shows an implementation of the multi-criteria multi-stakeholder approach for using the CAR method from [4], based on rankings and imprecise information. It is developed for group decisions for infrastructure policy making in Swedish municipalities.

To somewhat circumvent these problems, some approaches utilise imprecise importance information to determine criteria weights and sometimes values of alternatives.[1] Such methods were mostly assessed using case studies until [2] introduced a

---

[1] In this paper the full features of the large variety of elicitation techniques will not be discussed. For more exhaustive discussion refer to [6].

© Springer International Publishing AG 2017
D. Bajwa et al. (Eds.): GDN 2016, LNBIP 274, pp. 16–27, 2017.
DOI: 10.1007/978-3-319-52624-9_2

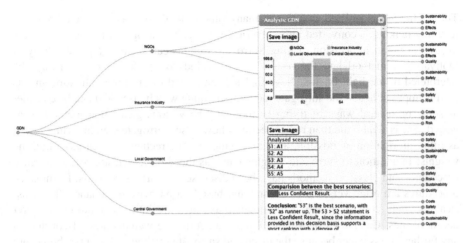

**Fig. 1.** The Group Decision tool Decision Wizard

process utilising systematic simulations from criteria orderings. The basic idea is to generate surrogate weights from a criteria ordering and investigate how well the results of using these surrogates match the ordering provided by the decision-maker, while trying to determine which the 'true' weights of the decision-makers are.

The methodology is of course vulnerable since the validation result is heavily dependent on the model we have of the decision-maker's mind set reflected in the distribution used for generating the weight vectors. To use these surrogate numbers, whichever way they are produced, they need to fulfil some robustness ideas, since we cannot for certain claim that we know exactly what decision-makers have in mind when stating such an ordering. This article discusses these robustness issues when translating orderings into surrogate numbers.

## 2  Rank Ordering Methods

Common for most elicitation methods is the assumption that all elicitation is made relative to a weight distribution held by the decision-maker.[2] One initial idea is to just skip the criteria elicitation and assign equal weights to every criterion. However, the information loss is then very large and it is most often worthwhile to at least rank the criteria, since rankings are (often) easier to provide than precise numbers. An ordering of the criteria is then achieved which can be handled in various ways. One such is to introduce so called surrogate weights, which are derived from the supposed ranking.

---

[2] For various cognitive and methodological aspects of imprecision in decision making, see e.g., [3] and others by the same authors.

This technique is utilised in [2] and many others. In these classes of methods, the resulting ranking is converted into numerical weights by surrogate functions. Needless to say, for practical decision making, surrogate weights can seem as a peculiar way of motivating a method and the results of these kinds of methods should always be interpreted in the light of this. Nevertheless, some kind of absolute validation in this field is impossible and the surrogate methods are quite widely used and can be consider as some of several ways of trying to motivate the various generation methods suggested. The crucial issue then rather becomes how to set surrogate weights while losing as little information as possible and preserving the "correctness" when assigning the weights. In addition to surrogate weights (such as those discussed above), dominance procedures and classical methods have been discussed in various contexts. Dominance procedures are often versions of outranking, based on pairwise dominance. The classical methods consist of the well-known maximax, maximin, and minimax regret decision rules. Categories of weights other than surrogate weights are not considered any further in this paper because the discussed surrogate methods, as will be shown, are very efficient.

## 3   The RS, RR and ROC Methods

In the literature, various surrogate weight methods have been suggested. [8] discusses rank sum (RS) weights and rank reciprocal (RR) weights, which are alternatives to the quite popular ROC (rank order centroid) from [1]. The rank sum is based on the idea that the rank order should be reflected directly in the weight. Assume a simplex $S_w$ generated by $w_1 > w_2 > ... > w_N$, $\Sigma w_i = 1$ and $0 \leq w_i$. We will, unless otherwise stated, henceforth presume that decision problems are modelled as simplexes $S_w$. Assign an ordinal number to each item ranked, starting with the highest ranked item as number 1. Denote the ranking number $i$ among $N$ items to rank. Thus, a larger weight is assigned to lower ranking numbers (Table 1).

$$w_i^{RS} = \frac{N+1-i}{\sum_{j=1}^{N}(N+1-j)} \tag{1}$$

Another idea discussed in [8] is rank reciprocal (RR) weights. They have a similar origin as the RS weights, but are based on the reciprocals (inverted numbers) of the rank order for each item ranked. These are obtained by assigning an ordinal number to each item ranked, starting with the highest ranked item as number 1. Thus, a larger weight is again assigned to lower ranking numbers (Table 2).

$$w_i^{RR} = \frac{1/i}{\sum_{j=1}^{N}\frac{1}{j}}, \text{ where } i \text{ and } j \text{ are as above.} \tag{2}$$

ROC weights are the centroid components of the simplex $S_w$. That is, ROC is a function based on the average of the corners in the polytope defined by the simplex $S_w = w_1 > w_2 > ... > w_N$, $\Sigma w_i = 1$, and $0 \leq w_i$.

$$w_i^{\text{ROC}} = 1/N \sum_{j=i}^{N} \frac{1}{j} \qquad (3)$$

In this way, it resembles RR more than RS but is, particularly for lower dimensions, more extreme than both in the sense of weight distribution, especially the largest and smallest weights (Table 3).

**Table 1.** The RS weights up to ten dimensions

| RS | i = 1 | i = 2 | i = 3 | i = 4 | i = 5 | i = 6 | i = 7 | i = 8 | i = 9 | i = 10 |
|---|---|---|---|---|---|---|---|---|---|---|
| N = 1 | 1.0000 | | | | | | | | | |
| N = 2 | 0.6667 | 0.3333 | | | | | | | | |
| N = 3 | 0.5000 | 0.3333 | 0.1667 | | | | | | | |
| N = 4 | 0.4000 | 0.3000 | 0.2000 | 0.1000 | | | | | | |
| N = 5 | 0.3333 | 0.2667 | 0.2000 | 0.1333 | 0.0667 | | | | | |
| N = 6 | 0.2857 | 0.2381 | 0.1905 | 0.1429 | 0.0952 | 0.0476 | | | | |
| N = 7 | 0.2500 | 0.2143 | 0.1786 | 0.1429 | 0.1071 | 0.0714 | 0.0357 | | | |
| N = 8 | 0.2222 | 0.1944 | 0.1667 | 0.1389 | 0.1111 | 0.0833 | 0.0556 | 0.0278 | | |
| N = 9 | 0.2000 | 0.1778 | 0.1556 | 0.1333 | 0.1111 | 0.0889 | 0.0667 | 0.0444 | 0.0222 | |
| N = 10 | 0.1818 | 0.1636 | 0.1455 | 0.1273 | 0.1091 | 0.0909 | 0.0727 | 0.0545 | 0.0364 | 0.0182 |

Of the three methods above, ROC has often been considered to be a quite reasonable candidate, despite that generated weights are sometimes perceived to be too sharp or discriminative, meaning that too large emphasis is put on the larger weights, i.e. on those criteria ranked highest up in the ranking order.

**Table 2.** The RR weights up to ten dimensions

| RR | i = 1 | i = 2 | i = 3 | i = 4 | i = 5 | i = 6 | i = 7 | i = 8 | i = 9 | i = 10 |
|---|---|---|---|---|---|---|---|---|---|---|
| N = 1 | 1.0000 | | | | | | | | | |
| N = 2 | 0.6667 | 0.3333 | | | | | | | | |
| N = 3 | 0.5455 | 0.2727 | 0.1818 | | | | | | | |
| N = 4 | 0.4800 | 0.2400 | 0.1600 | 0.1200 | | | | | | |
| N = 5 | 0.4380 | 0.2190 | 0.1460 | 0.1095 | 0.0876 | | | | | |
| N = 6 | 0.4082 | 0.2041 | 0.1361 | 0.1020 | 0.0816 | 0.0680 | | | | |
| N = 7 | 0.3857 | 0.1928 | 0.1286 | 0.0964 | 0.0771 | 0.0643 | 0.0551 | | | |
| N = 8 | 0.3679 | 0.1840 | 0.1226 | 0.0920 | 0.0736 | 0.0613 | 0.0526 | 0.0460 | | |
| N = 9 | 0.3535 | 0.1767 | 0.1178 | 0.0884 | 0.0707 | 0.0589 | 0.0505 | 0.0442 | 0.0393 | |
| N = 10 | 0.3414 | 0.1707 | 0.1138 | 0.0854 | 0.0683 | 0.0569 | 0.0488 | 0.0427 | 0.0379 | 0.0341 |

When comparing the centroid weights of the ROC, RS, and RR, we can see that there are significant differences. To begin with, ROC is compared to RR in Table 4. RR shows a rather large similarity to ROC in terms of generating function. While it is true that ROC puts larger emphasis on the higher ranked criteria than the other methods for lower $N$, RR quickly takes over as the most heavy emphasizer of the highest ranked criterion from $N = 6$ upwards.

**Table 3.** The ROC weights up to 10 dimensions

| ROC | $i = 1$ | $i = 2$ | $i = 3$ | $i = 4$ | $i = 5$ | $i = 6$ | $i = 7$ | $i = 8$ | $i = 9$ | $i = 10$ |
|---|---|---|---|---|---|---|---|---|---|---|
| $N = 1$ | 1.0000 | | | | | | | | | |
| $N = 2$ | 0.7500 | 0.2500 | | | | | | | | |
| $N = 3$ | 0.6111 | 0.2778 | 0.1111 | | | | | | | |
| $N = 4$ | 0.5208 | 0.2708 | 0.1458 | 0.0625 | | | | | | |
| $N = 5$ | 0.4567 | 0.2567 | 0.1567 | 0.0900 | 0.0400 | | | | | |
| $N = 6$ | 0.4083 | 0.2417 | 0.1583 | 0.1028 | 0.0611 | 0.0278 | | | | |
| $N = 7$ | 0.3704 | 0.2276 | 0.1561 | 0.1085 | 0.0728 | 0.0442 | 0.0204 | | | |
| $N = 8$ | 0.3397 | 0.2147 | 0.1522 | 0.1106 | 0.0793 | 0.0543 | 0.0335 | 0.0156 | | |
| $N = 9$ | 0.3143 | 0.2032 | 0.1477 | 0.1106 | 0.0828 | 0.0606 | 0.0421 | 0.0262 | 0.0123 | |
| $N = 10$ | 0.2929 | 0.1929 | 0.1429 | 0.1096 | 0.0846 | 0.0646 | 0.0479 | 0.0336 | 0.0211 | 0.0100 |

**Table 4.** ROC compared to RR

| ROC–RR | $i = 1$ | $i = 2$ | $i = 3$ | $i = 4$ | $i = 5$ | $i = 6$ | $i = 7$ | $i = 8$ | $i = 9$ | $i = 10$ |
|---|---|---|---|---|---|---|---|---|---|---|
| $N = 1$ | 0.0000 | | | | | | | | | |
| $N = 2$ | 0.0833 | –0.0833 | | | | | | | | |
| $N = 3$ | 0.0657 | 0.0051 | –0.0707 | | | | | | | |
| $N = 4$ | 0.0408 | 0.0308 | –0.0142 | –0.0575 | | | | | | |
| $N = 5$ | 0.0187 | 0.0377 | 0.0107 | –0.0195 | –0.0476 | | | | | |
| $N = 6$ | 0.0002 | 0.0376 | 0.0223 | 0.0007 | –0.0205 | –0.0402 | | | | |
| $N = 7$ | –0.0153 | 0.0347 | 0.0276 | 0.0121 | –0.0043 | –0.0201 | –0.0347 | | | |
| $N = 8$ | –0.0282 | 0.0308 | 0.0296 | 0.0186 | 0.0057 | –0.0070 | –0.0191 | –0.0304 | | |
| $N = 9$ | –0.0392 | 0.0265 | 0.0298 | 0.0223 | 0.0122 | 0.0017 | –0.0084 | –0.0180 | –0.0269 | |
| $N = 10$ | –0.0485 | 0.0222 | 0.0291 | 0.0242 | 0.0163 | 0.0077 | –0.0009 | –0.0091 | –0.0168 | –0.0241 |

RR also emphasizes the lowest ranked criteria, making it the weighing method that puts the most emphasis on the end-points of the ranking. The differences in the table (and in some of the subsequent ones) might not appear as too dramatic, but better precision of a method, even if it by some smaller percentage, is a way of fine tuning these originally quite rough methods based of a kind of statistical validity.

**Table 5.** ROC compared to RS

| ROC–RS | i = 1 | i = 2 | i = 3 | i = 4 | i = 5 | i = 6 | i = 7 | i = 8 | i = 9 | i = 10 |
|--------|-------|-------|-------|-------|-------|-------|-------|-------|-------|--------|
| N = 1  | 0.0000 |        |        |        |        |        |        |        |        |         |
| N = 2  | 0.0833 | −0.0833 |        |        |        |        |        |        |        |         |
| N = 3  | 0.1111 | −0.0556 | −0.0556 |        |        |        |        |        |        |         |
| N = 4  | 0.1208 | −0.0292 | −0.0542 | −0.0375 |        |        |        |        |        |         |
| N = 5  | 0.1233 | −0.0100 | −0.0433 | −0.0433 | −0.0267 |        |        |        |        |         |
| N = 6  | 0.1226 | 0.0036 | −0.0321 | −0.0401 | −0.0341 | −0.0198 |        |        |        |         |
| N = 7  | 0.1204 | 0.0133 | −0.0224 | −0.0344 | −0.0344 | −0.0272 | −0.0153 |        |        |         |
| N = 8  | 0.1175 | 0.0203 | −0.0144 | −0.0283 | −0.0318 | −0.0290 | −0.0221 | −0.0122 |        |         |
| N = 9  | 0.1143 | 0.0254 | −0.0079 | −0.0227 | −0.0283 | −0.0283 | −0.0246 | −0.0182 | −0.0099 |         |
| N = 10 | 0.1111 | 0.0293 | −0.0026 | −0.0177 | −0.0245 | −0.0263 | −0.0248 | −0.0209 | −0.0153 | −0.0082 |

Comparing ROC to RS in Table 5, it is evident that ROC emphasizes the higher ranked criteria at the expense of the middle and lower ranked ones. For the middle ranked criteria, this is the opposite of the RR method when compared to ROC.

## 4    The SR Method

Not surprisingly, ROC, RS and RR perform well only for specific assumptions on decision-maker assignments of criteria weight preferences and since these weight models are in a sense opposites, it is interesting to see how extreme behaviours can be reduced. A natural candidate for this could be a linear combination of RS and RR. Since we have no reasons to expect anything else, we can, e.g., balance them equally in an additive combination of the Sum and the Reciprocal weight function that we call the SR weight method.

$$
w_i^{SR} = \frac{1/i + \frac{N+1-i}{N}}{\sum_{j=1}^{N} \left(1/j + \frac{N+1-j}{N}\right)}, \text{ i and j as above.} \tag{4}
$$

Of course, other combinations of these would be thinkable, but the important observation is achieved by comparing SR with the others. The actual mix between the methods would affect the result according to its proportions. The details there are not crucial for our main point that all these results are a sensitive product of the underlying assumptions regarding the mind-settings of decision-makers. If this nevertheless would be important, the reasonably proportions must be elicited and fine-tuned with respect to the individual in question. A reasonable meaning of such a procedure escapes us and is in any case beyond the scope of this article, so the tables below show the equally proportional SR weights and its behaviour in relation to ROC and RS. Table 6 shows the weights $w_i^{SR}$ for different numbers of criteria up to $N = 10$.

**Table 6.** The weights for SR

| ROC–RS | i = 1 | i = 2 | i = 3 | i = 4 | i = 5 | i = 6 | i = 7 | i = 8 | i = 9 | i = 10 |
|---|---|---|---|---|---|---|---|---|---|---|
| N = 1 | 0.0000 | | | | | | | | | |
| N = 2 | 0.0833 | –0.0833 | | | | | | | | |
| N = 3 | 0.1111 | –0.0556 | –0.0556 | | | | | | | |
| N = 4 | 0.1208 | –0.0292 | –0.0542 | –0.0375 | | | | | | |
| N = 5 | 0.1233 | –0.0100 | –0.0433 | –0.0433 | –0.0267 | | | | | |
| N = 6 | 0.1226 | 0.0036 | –0.0321 | –0.0401 | –0.0341 | –0.0198 | | | | |
| N = 7 | 0.1204 | 0.0133 | –0.0224 | –0.0344 | –0.0344 | –0.0272 | –0.0153 | | | |
| N = 8 | 0.1175 | 0.0203 | –0.0144 | –0.0283 | –0.0318 | –0.0290 | –0.0221 | –0.0122 | | |
| N = 9 | 0.1143 | 0.0254 | –0.0079 | –0.0227 | –0.0283 | –0.0283 | –0.0246 | –0.0182 | –0.0099 | |
| N = 10 | 0.1111 | 0.0293 | –0.0026 | –0.0177 | –0.0245 | –0.0263 | –0.0248 | –0.0209 | –0.0153 | –0.0082 |

Table 7 demonstrates how the weights for this SR combination compare to ROC weights (RR is similar to ROC in this respect). Similarly, Table 8 demonstrates how the weights for this SR combination compare to RS weights. From Tables 7 and 8 we can, as expected, see that SR does indeed constitute a compromise that tries to compensate for shortcomings in the other methods but does not deviate too much from any of them.

**Table 7.** SR compared to ROC weights

| SR–ROC | i = 1 | i = 2 | i = 3 | i = 4 | i = 5 | i = 6 | i = 7 | i = 8 | i = 9 | i = 10 |
|---|---|---|---|---|---|---|---|---|---|---|
| N = 1 | 0.0000 | | | | | | | | | |
| N = 2 | –0.0833 | 0.0833 | | | | | | | | |
| N = 3 | –0.0894 | 0.0266 | 0.0628 | | | | | | | |
| N = 4 | –0.0845 | 0.0019 | 0.0360 | 0.0466 | | | | | | |
| N = 5 | –0.0781 | –0.0106 | 0.0200 | 0.0330 | 0.0357 | | | | | |
| N = 6 | –0.0722 | –0.0176 | 0.0097 | 0.0233 | 0.0285 | 0.0282 | | | | |
| N = 7 | –0.0670 | –0.0217 | 0.0028 | 0.0161 | 0.0226 | 0.0244 | 0.0229 | | | |
| N = 8 | –0.0626 | –0.0242 | –0.0021 | 0.0107 | 0.0177 | 0.0207 | 0.0209 | 0.0190 | | |
| N = 9 | –0.0589 | –0.0258 | –0.0057 | 0.0065 | 0.0137 | 0.0174 | 0.0187 | 0.0181 | 0.0160 | |
| N = 10 | –0.0556 | –0.0268 | –0.0084 | 0.0031 | 0.0103 | 0.0145 | 0.0165 | 0.0168 | 0.0158 | 0.0137 |

**Table 8.** SR compared to RS weights

| SR–RS | $i = 1$ | $i = 2$ | $i = 3$ | $i = 4$ | $i = 5$ | $i = 6$ | $i = 7$ | $i = 8$ | $i = 9$ | $i = 10$ |
|---|---|---|---|---|---|---|---|---|---|---|
| $N = 1$ | 0.0000 | | | | | | | | | |
| $N = 2$ | 0.0000 | 0.0000 | | | | | | | | |
| $N = 3$ | 0.0217 | −0.0290 | 0.0072 | | | | | | | |
| $N = 4$ | 0.0364 | −0.0273 | −0.0182 | 0.0091 | | | | | | |
| $N = 5$ | 0.0452 | −0.0206 | −0.0233 | −0.0103 | 0.0090 | | | | | |
| $N = 6$ | 0.0504 | −0.0140 | −0.0224 | −0.0168 | −0.0056 | 0.0084 | | | | |
| $N = 7$ | 0.0534 | −0.0084 | −0.0197 | −0.0183 | −0.0118 | −0.0028 | 0.0076 | | | |
| $N = 8$ | 0.0549 | −0.0039 | −0.0166 | −0.0177 | −0.0141 | −0.0083 | −0.0011 | 0.0069 | | |
| $N = 9$ | 0.0555 | −0.0004 | −0.0136 | −0.0162 | −0.0146 | −0.0108 | −0.0058 | −0.0001 | 0.0062 | |
| $N = 10$ | 0.0555 | 0.0025 | −0.0110 | −0.0146 | −0.0142 | −0.0118 | −0.0083 | −0.0041 | 0.0005 | 0.0055 |

## 5   The EW, SIMOS and RE Methods

In the 1970s, the quite naïve idea of equal weights (EW) gained some recognition. The hope was that, given that elicitation is a hard problem, equal weights would perform as well as any other set of weights. As a generalization to RS as previously discussed, a rank exponent weight method was introduced by [8]. In the RS formula, introduce the exponent $z$ to yield the rank exponent (RE) weights as

$$w_i^{\text{RE}} = \frac{(N+1-i)^z}{\sum_{j=1}^{N}(N+1-j)^z}. \tag{5}$$

Thus, $z$ mediates between equal weights and the RS weights and for $z = 0$, this becomes equal weights and for $z = 1$ it becomes RS weights. Thus, for $0 < z < 1$ it is the exponential combination of equal and RS weights. Beyond $z = 1$ it becomes a more extreme weighting scheme. This makes the $z$ parameter of RE a bit hard to estimate and potentially less suitable in real-life decisions.

Another type of method is the SIMOS method, which has gained some interest in these contexts. It was proposed in [7] with the purpose of providing decision makers with a simple method, not requiring any former familiarity with decision analytical techniques and can easily express criteria hierarchies introducing some cardinality as well, if needed. It has been applied in a multitude of contexts and seems to have been comparatively well received by real-life decision-makers. The SIMOS method has, however, been criticised for not being robust when the preferences are changed and it has further some contra-intuitive features. [5] suggested a revised version, but introduces the severely complicating factor to correctly estimate a reliable and robust proportional factor $z$ between the most and least important criteria. In this study, the

Simos weights are used ordinally by not using the blank cards or equal weights. The $z$ factor is estimated at $N + 1$, with $N$ being the number of criteria as usual.

## 6    Assessing Models for Surrogate Weights

Simulation studies have become a kind of de facto standard for comparing multi-criteria weights. The assumption is that there exist a set of 'true' weights in the decision-maker's mind which are inaccessible in its pure form by any elicitation method. The modelling assumptions regarding decision-makers' mind-sets above are then inherent in the generation of decision problem vectors by a random generator. When following an $N$–1 DoF model, a vector is generated in which the components sum to 100%. This simulation is based on a homogenous $N$-variate Dirichlet distribution generator. When following an $N$ DoF model, a vector is generated without an initial joint restriction, only keeping components within [0%, 100%] implying $N$ degrees of freedom, where these components subsequently are normalised so that their sum is 100%. We call the $N$–1 DoF model type of generator an *$N-1$-generator* and the $N$ DoF model type an *$N$-generator*. Depending on how we model the decision-maker's weight assessments, the results then become very different: ROC weights in $N$ dimensions coincide with the mass point for the vectors of the $N-1$-generator over the polytope $S_w$. Similarly, RS weights are very close to the mass point of an $N$-generator over a polytope. In reality, though, we cannot know whether a specific decision-maker (or decision-makers in general) adhere more to $N-1$ or $N$ DoF representations of their knowledge. Both as individuals and as a group, they might use either or be anywhere in between. A, in a reasonable sense, *robust* rank ordering mechanism must employ a surrogate weight function that at least handles both types of conceptualisation and anything in between.

The simulations were carried out with a varying number of criteria and alternatives. There were four numbers of criteria $N = \{3, 6, 9, 12\}$ and five numbers of alternatives $M = \{3, 6, 9, 12, 15\}$ creating a total of 20 simulation scenarios. Each scenario was run 10 times, each time with 10,000 trials, yielding a total of 2,000,000 decision situations generated. For this simulation, an $N$-variate joint Dirichlet distribution was employed to generate the random weight vectors for the $N$–1 DoF simulations and a standard round-robin normalised random weight generator for the $N$ DoF simulations. Unscaled value vectors were generated uniformly, and no significant differences were observed with other value distributions.

The results of the simulations are shown in the tables below, where we show a subset of the results with chosen pairs $(N, M)$. There were three measures of success.[3] The first is the hit ratio as in the previous studies ("winner"), the number of times the highest evaluated alternative using a particular method coincides with the true highest alternative. The second is the matching of the three highest ranked alternatives ("podium"), the number of times the three highest evaluated alternatives using a

---

[3] Kendall's tau was also computed and it does not deviate from the other findings in the tables. Thus, it is not shown in the tables.

particular method all coincide with the true three highest alternatives. This means that the internal order between these three alternatives is uninteresting. A third set generated is the matching of all ranked alternatives ("overall"), the number of times all evaluated alternatives using a particular method coincide with the true ranking of the alternatives. All three sets correlated strongly with each other and the latter two are not shown in this paper. The tables show the winner frequency for the six methods ROC, RE, RR, SIMOS, SR, and EW utilising the simulation methods $N-1$ DoF, $N$ DoF and a 50% combination of $N-1$ DoF and $N$ DoF. All hit ratios in all tables are given in per cent and are mean values of the 10 scenario runs. The standard deviations between sets of 10 runs were around 0.2–0.3 per cent. In Table 9, using an $N-1$-generator, it can be seen that ROC not surprisingly outperforms the others, when looking at the winner, but with SR close behind. EW, likewise not surprisingly, is performing much worse than all the others.

**Table 9.** Using an N–1-generator, it can be seen that ROC outperforms the others

| N–1 DoF | | ROC | RE | RR | Simos | SR | EW |
|---|---|---|---|---|---|---|---|
| 3 criteria | 3 alternatives | 90.2 | 88.9 | 89.5 | 88.1 | 89.3 | 72.9 |
| 3 criteria | 15 alternatives | 79.1 | 77.2 | 76.5 | 76.2 | 76.9 | 56.5 |
| 6 criteria | 6 alternatives | 84.8 | 80.6 | 82.7 | 79.6 | 83.1 | 57.5 |
| 6 criteria | 12 alternatives | 81.3 | 76.9 | 78.2 | 75.5 | 78.9 | 50.0 |
| 9 criteria | 9 alternatives | 83.5 | 77.6 | 79.5 | 76.2 | 81.2 | 50.1 |
| 12 criteria | 6 alternatives | 86.4 | 78.5 | 80.8 | 77.1 | 84.1 | 54.2 |
| 12 criteria | 12 alternatives | 83.4 | 74.2 | 76.8 | 72.5 | 80.2 | 45.7 |

In Table 10 the frequencies have changed according to expectation since we employ a model with $N$ degrees of freedom. Now the SIMOS model behaves very similar to RE and, as expected, outperforms the others, in particular when it comes to a larger number of criteria and alternatives. In Table 11, the $N$ and $N-1$ DoF models are combined with equal emphasis on both. Now, we can see that in total RE and SR perform the best.

**Table 10.** For N degrees of freedom, RE and Simos are top of the form

| N DoF | | ROC | RE | RR | Simos | SR | EW |
|---|---|---|---|---|---|---|---|
| 3 criteria | 3 alternatives | 87.3 | 89.3 | 88.3 | 89.2 | 89.1 | 78.3 |
| 3 criteria | 15 alternatives | 77.9 | 81.3 | 79.1 | 81.4 | 80.6 | 65.6 |
| 6 criteria | 6 alternatives | 80.1 | 87.3 | 78.1 | 87.4 | 85.1 | 67.4 |
| 6 criteria | 12 alternatives | 76.4 | 84.2 | 74.3 | 84.3 | 82.0 | 60.9 |
| 9 criteria | 9 alternatives | 76.3 | 87.0 | 69.8 | 87.2 | 83.0 | 62.3 |
| 12 criteria | 6 alternatives | 77.5 | 90.0 | 67.8 | 90.2 | 84.6 | 66.4 |
| 12 criteria | 12 alternatives | 73.4 | 87.4 | 63.1 | 87.7 | 81.7 | 58.7 |

**Table 11.** Winners when considering both DoF models

| Combined | | ROC | RE | RR | Simos | SR | EW |
|---|---|---|---|---|---|---|---|
| 3 criteria | 3 alternatives | 88.8 | 89.1 | 88.9 | 88.7 | 89.2 | 75.6 |
| 3 criteria | 15 alternatives | 78.5 | 79.3 | 77.8 | 78.8 | 78.8 | 61.1 |
| 6 criteria | 6 alternatives | 82.5 | 84.0 | 80.4 | 83.5 | 84.1 | 62.5 |
| 6 criteria | 12 alternatives | 78.9 | 80.6 | 76.3 | 79.9 | 80.5 | 55.5 |
| 9 criteria | 9 alternatives | 79.9 | 82.3 | 74.7 | 81.7 | 82.1 | 56.2 |
| 12 criteria | 6 alternatives | 82.0 | 84.3 | 74.3 | 83.7 | 84.4 | 60.3 |
| 12 criteria | 12 alternatives | 78.4 | 80.8 | 70.0 | 80.1 | 81.0 | 52.2 |

Since we are looking for a surrogate model with both precision and robustness, we turn our attention to the spread between the methods' results under the two varying decision-maker assumptions regarding degrees of freedom when producing weights. Table 12 shows the spread as the absolute value of the difference between the frequencies (hit ratios) under the $N-1$ DoF model and the $N$ DoF model.

**Table 12.** Spread as the absolute value of the difference between the frequencies (hit ratios) under the N–1 DoF model and the N DoF model for winners

| Spread | | ROC | RE | RR | Simos | SR | EW |
|---|---|---|---|---|---|---|---|
| 3 criteria | 3 alternatives | 2.9 | 0.4 | 1.2 | 1.1 | 0.2 | 5.4 |
| 3 criteria | 15 alternatives | 1.2 | 4.1 | 2.6 | 5.2 | 3.7 | 9.1 |
| 6 criteria | 6 alternatives | 4.7 | 6.7 | 4.6 | 7.8 | 2.0 | 9.9 |
| 6 criteria | 12 alternatives | 4.9 | 7.3 | 3.9 | 8.8 | 3.1 | 10.9 |
| 9 criteria | 9 alternatives | 7.2 | 9.4 | 9.7 | 11.0 | 1.8 | 12.2 |
| 12 criteria | 6 alternatives | 8.9 | 11.5 | 13.0 | 13.1 | 0.5 | 12.2 |
| 12 criteria | 12 alternatives | 10.0 | 13.2 | 13.7 | 15.2 | 1.5 | 13.0 |

## 7   Concluding Remarks

The aim of this study has been to find reasonably robust multi-criteria weights that are able to cover a broad set of decision situations, but at the same time have a reasonably simple semantic regarding how they are generated. In this paper, we have considered $N = \{3, 6, 9, 12, 15\}$ and $M = \{3, 6, 9, 12\}$ and to summarise the analysis, we look at the average hit rate in per cent ("mean correct") over all the pairs $(N, M)$ that we have reported in the tables above. Table 13 shows the conclusion of the performances. RE heads the table with SR not far behind. Further in Table 13, we can see the mean square spread between the different DoF. RE and SR are the best candidates when it comes to the mean value, followed by Simos. However, the robustness in the sense of mean square variation is significant. In that respect SR is clearly the most superior, rendering it the top ranked position. Since we appreciate both precision and robustness, the final

**Table 13.** Performance averages and mean square for the six methods

| Conclusion | ROC | RE | RR | Simos | SR | EW |
|---|---|---|---|---|---|---|
| Mean correct (Table 11) | 81.3 | 82.9 | 77.5 | 82.3 | 82.8 | 60.5 |
| Rank from mean correct | 4 | 1 | 5 | 3 | 2 | 6 |
| Mean square spread (Table 12) | 6.4 | 8.5 | 8.4 | 9.9 | 2.2 | 10.7 |
| Rank from mean sq spread | 2 | 4 | 3 | 5 | 1 | 6 |
| Final score | 74.9 | 74.4 | 69.1 | 72.4 | 80.7 | 49.8 |
| Rank from final score | 2 | 3 | 5 | 4 | 1 | 6 |

score determining the most suitable surrogate weight method is the difference between the mean value of the "winners" and the mean value of the squared spread.

It is clear that SR is the preferred surrogate method of those investigated. In a further study, it would be interesting to study the RE method with a varying set of exponents to see if primarily the stability can be improved, in that case making it a candidate for real-life decision making tools.

# References

1. Barron, F.H.: Selecting a best multiattribute alternative with partial information about attribute weights. Acta Psychol. **80**(1–3), 91–103 (1992)
2. Barron, F., Barrett, B.: Decision quality using ranked attribute weights. Manag. Sci. **42**(11), 1515–1523 (1996)
3. Danielson, M., Ekenberg, L., Larsson, A.: Distribution of belief in decision trees. Int. J. Approx. Reason. **46**(2), 387–407 (2007)
4. Danielson, M., Ekenberg, L.: The CAR method for using preference strength in multi-criteria decision making. Group Decis. Negot. **25**, 775–797 (2015)
5. Figueira, J., Roy, B.: Determining the weights of criteria in the ELECTRE type methods with a revised Simos' procedure. Eur. J. Oper. Res. **139**, 317–326 (2002)
6. Riabacke, M., Danielson, M., Ekenberg, L.: State-of-the-art in prescriptive weight elicitation. Adv. Decis. Sci. Article ID 276584, p. 24 (2012). doi:10.1155/2012/276584
7. Simos, J.: L'evaluation environnementale: Un processus cognitif neegociee. Theese de doctorat, DGF-EPFL, Lausanne (1990)
8. Stillwell, W., Seaver, D., Edwards, W.: A comparison of weight approximation techniques in multiattribute utility decision making. Organ. Behav. Hum. Perform. **28**(1), 62–77 (1981)

# Group Decision Making

# Multi-criteria Group Decision Making
# with Private and Shared Criteria:
# An Experiment

Pascale Zaraté[1(✉)] , Guy Camilleri[2] , and D. Marc Kilgour[3]

[1] IRIT, Toulouse Capitole University, Toulouse, France
pascale.zarate@irit.fr
[2] IRIT, Toulouse University, Toulouse, France
[3] Wilfrid Laurier University, Waterloo, Canada

**Abstract.** Collective decision processes remain a common management approach in most organizations. In such processes, it seems important to offer participants the opportunity to confront the differences in their points of view. To this end, cognitive and technical tools are required that facilitate the sharing of individuals' reasoning and preferences, but at the same time allow them to keep some information and attitudes to themselves. The aim of our study is to assess whether, in the multi-criteria approach to problem structuring, decision-makers can be comfortable using shared criteria in addition to private criteria. For this purpose, an exploratory experiment with student subjects was conducted using the Group Decision Support System, GRUS.

**Keywords:** GDSS · Multi-criteria group decision making · Private criteria · Public criteria

## 1 Introduction

In large organizations, the vast majority of decisions are taken after extensive consultations involving numerous individuals, rather than by individual decision makers [1]. According to Smoliar and Sprague [2], decision making in organizations generally involves the interaction of several actors. This interaction includes communication of information, but its main aim is to enable the decision makers to come to a shared understanding, thereby assisting them at achieving a coordinated solution to the problem at hand.

The process of group decision making has been analyzed from a number of perspectives. Recently, Zaraté [3] suggested that the increasing complexity of organizations, and the use of Information and Communication Technologies to support them, require decision processes to be modified. On the organizational level, processes now involve more actors with greater amounts of responsibility, while at the individual level, decision makers face more demands on their cognitive processes; they not only face greater quantities of information, but must also make sense of it rapidly. A new kind of Cooperative Decision Process is now needed.

© Springer International Publishing AG 2017
D. Bajwa et al. (Eds.): GDN 2016, LNBIP 274, pp. 31–42, 2017.
DOI: 10.1007/978-3-319-52624-9_3

To support a group engaged in decision making, Macharis et al. [4] introduced a methodology based on the Multiple Criteria paradigm through the PROMETHEE methodology. They propose that each decision maker create his or her own performance matrix by determining his or her own individual values. Then a global evaluation of each alternative is performed using a weighted sum aggregation technique. Decision makers' weights may be equal or different. One benefit of this structure is the ability to conduct a stakeholder-level sensitivity analysis. Nevertheless, the proposed system does not permit the decision makers to share their preferences with others, or to co-build a decision.

In a collective decision framework, decision makers must balance their own attitudes and preferences with the goal of building common preferences and consensus within the group. The purpose of this paper is to investigate whether decision makers can feel comfortable making decisions that integrate common and individual preferences. We conducted an experiment based on the multi-criteria Group Decision Support System, GRoUp Support (GRUS). This experiment is an exploratory research. The aim was to explore parameters that highlight the advantages and disadvantages of a multi-criteria Group Decision process. In practice, are the advantages noticeable, and under what conditions? We also wish to assess whether participants can perceive the advantages of the group multi-criteria approach to decision-making.

This paper is structured as follows. The GRUS system is described in Sect. 2. Then the Research Questions of our study are set out in Sect. 3, and the experiment is described. Section 4 gives the experimental results and analyzes them, and Sect. 5 discusses the implications of the results for the hypotheses in Sect. 3. Finally, in Sect. 6, we give concluding remarks and perspectives on this work.

## 2    GRoUp Support: The GRUS System

GRUS, a free web platform available at http://www.irit.fr/GRUS, can support several kinds of meetings: synchronous, asynchronous, distributed or face-to-face. Sometimes, additional components are required; for example, a distributed/asynchronous decision process is managed by a facilitator as a classical project, with agenda, deadlines, etc. GRUS is protected by a login and a password, available from the authors upon request.

GRUS is designed as a toolbox and implemented in Grails, a framework based on Groovy, a very high level language similar to Python or Ruby. Groovy can be compiled to a Java Virtual Machine bytecode and can interoperate with other java codes or libraries. For more details, see [5].

GRUS is designed to support different types of users, including designers of collaborative tools (application developers), designers of collaborative processes (collaboration engineers), session facilitators and decision makers (users of GRUS). It offers the basic services commonly available in Group Decision Support, such as definition/design of a group decision process, both static and dynamic way; management (add, modify, delete, etc.) of collaborative tools; management of automatic reporting in PDF format, etc. It is conceived as a toolbox for Collaborative Decision Processes, and includes a Brainstorming tool, a Clustering tool, multi-criteria Analysis, a Voting tool, a Consensus tool, and a Reporting tool.

Using the multi-criteria tools, users can define criteria and alternatives, and give their own assessments of the performance of each alternative on each criterion, creating a performance matrix. Each assessment is on a scale from 0 to 20. Decision makers may also indicate their own (individual) preferences for the weight of each criterion.

In order to enter his or her own preferences, a decision maker must enter a so-called suitability function, including an indifference threshold. In this way, the user's interpretation of each criterion can be taken into account. Finally, each decision maker's assessment of the dependencies between pairs of criteria can also be entered. Again, these dependencies are marked by each decision maker on a scale from 0 to 20.

Two aggregation techniques are implemented in the GRUS system. The first aggregation methodology is the weighted sum [6], which ignores any possible dependencies among criteria. The second aggregation methodology is the Choquet Integral [7], which does reflect dependencies among pairs of criteria.

## 3 The Experiment

### 3.1 Research Questions

Ideally, a group decision-making process includes much sharing of information as participants develop a common preference, leading to a good decision. Thanks to discussion, a better knowledge of the alternatives and the matches between preferences and alternatives is then possible. But participants may often simply announce their preferred alternative, without providing arguments about its appropriateness to solve the problem at hand. If so, the decision process does not contribute to any deeper understanding of the problem, and the decision does not benefit from being taken by a group [8]. Moreover, open sharing and extensive discussion are seldom practicable, first because participants have personal information or considerations that they do not (for reasons of strategy or privacy) reveal. Secondly, some aspects of individuals' preferences may not be crystal clear to themselves.

Thus, it is common that the result of a group decision-making process is based on a mix of objective and subjective reasoning. Recognizing this feature, Sibertin-Blanc and Zaraté [9] proposed a methodology distinguishing collective criteria and individual criteria, defined as follows:

- A criterion is *collective* if the group participants agree not only on its relevance, but also on the score of each alternative on this criterion;
- A criterion is *individual* if it is considered relevant by one (or several, but not all) participant, or if the participants do not agree on the scores of alternatives on this criterion.

The collective criteria constitute to the objective part of the group's assessment, while individual criteria are its subjective part.

*Research Question 1: In the design of a collaborative decision process, participants benefit from the availability of both private and common criteria.*

In order to guarantee group cohesion and consistency, it is necessary to find a balance between the individual part of the problem, i.e. the private criteria, and the collective part, i.e. the common criteria. On the other hand, it seems to us that if the number of private criteria is greater than the number of common criteria the decision is not really a group decision, but rather a collection of individual decisions.

*Research Question 2: In a collaborative decision making process, the number of private criteria should at most equal the number of common criteria.*

The role of Group Decision Support Systems is to support collaborative decision processes. Often a GDSS requires group facilitation, defined as a process in which a person acceptable to all members of the group intervenes to improve the group's identification and solution of problems, and the decisions it makes [10]. Facilitation is a dynamic process that involves managing relationships between people, tasks, and technology, as well as structuring tasks and contributing to the effective accomplishment of the intended outcomes.

Ackermann and Eden [11] found that facilitation helped groups to contribute freely to the discussion, to concentrate on the task, to sustain interest and motivation to solve the problem, to review progress and to address complicated issues rather than ignore them. A further task of facilitation is to engage the group in creativity and problem formulation techniques and to help it bring structure to the issues it faces [12]. Facilitators attend to the process of decision making, while the decision makers concentrate on the issues themselves.

Can facilitation be automatic? It has been argued that automatic facilitation enriches a GDSS as it guides decision makers toward successful structuring and execution of the decision making process [13]. According to [14], an electronic facilitator should execute four functions: (1) provide technical support by initiating and terminating specific software tools; (2) chair the meeting, maintaining and updating the agenda; (3) assist in agenda planning; and finally (4) provide organizational continuity, setting rules and maintaining an organizational repository.

Nevertheless, it seems difficult to program a process leading to insightful and creative decisions. Can a GDSS work well without a human facilitator?

*Research Question 3: GDSS use remains difficult without a human facilitator.*

## 3.2   Description

The experiment was conducted at Toulouse Capitole 1 University. One Master-level computer science class comprising 14 students was selected to participate. Three groups were created, including 4, 4 and 6 participants respectively. Each group worked independently, in a one-session meeting of 90 min. After the decision process, each participant responded to a questionnaire composed of seven questions, five about the common versus private criteria (Research Questions 1 and 2) and two about facilitation (Research Question 3).

The case-study decision problem was presented to each group is described below.

"You are member of the Administrative Committee of the Play-On-Line Company. This company develops Software Games. It includes 150 collaborators represented as follows:

- 80% Computer Engineers;
- 15% Business Staff;
- 5% Administrative Staff.

During a previous meeting, the Board decided to buy new mobile phones for all collaborators (the whole company). The use of the phones will not be the same for the three groups of collaborators. The computer engineers need to test the software as it is developed, on every operating system (Android, iPhone, etc.); the business staff will demonstrate the software to potential clients (and need large screens, for example). The administrative needs are simpler and more basic, such as communication (email, text, telephone, etc.).

The aim of today's meeting is to make together a decision about the best solution for Play-on-Line. The budget is strictly limited, so costs must be minimized. In order to satisfy the requirements of all stakeholders, your group must think up several solutions or scenarios but you must remember that company survival, from a financial point of view, is mandatory.

You can, for example, decide to buy the same Smartphones for everybody, or you can buy different models of smartphones for different collaborators, including some to be used only for testing. The technical characteristics and prices of five preselected Smartphones are given in the attached documents.

First, you have to define the set of criteria to be used (4–5) to solve this problem, and identify several alternatives (4–5). One alternative is defined as a combination of several smartphones, for example: 80% of Type A + 20% of Type B. You will be guided by the facilitator, and then you will enter in the GRUS system your own preferences used for calculating the group decision".

Each group was required to find 4–5 criteria and 4–5 alternatives, in order to restrict each session to 90 min. If the number of criteria and alternatives were decided by the group, we would not have been able to control the time of each session.

Using the GRUS system under the guidance of a human facilitator, the following process was applied:

- Brainstorming: Criteria and Alternatives are generated electronically. Each decision maker expresses himself or herself anonymously.
- Clustering: The number of criteria and the number of alternatives are reduced to 4–5. This step is conducted by the facilitator orally. Decision makers express themselves aloud in order to categorize all the ideas. The facilitator then categorizes the criteria and alternatives until the target numbers of criteria and alternatives (4–5) are achieved.
- Multi-criteria Evaluation: Decision makers give their own preferences on a scale to 0 to 20 for the performance of each alternative on each criterion. They also decide the weight of each criterion and the way that the criterion is to be interpreted (the suitability function – essentially a threshold score below which performance differences were ignored). Pairwise dependencies among criteria were also specified.

- Direct Vote: For this step the facilitator shows the results of the Multi-Criteria analysis. This result integrates all preferences given by all users and the results obtained by two ranking techniques: weighted sum and Choquet Integral, producing two total orders. A discussion is then initiated by the facilitator in order to classify all alternatives into three categories: A (Kept), C (Not Kept), and B (Feasible but uncertain). A "Kept" alternative will be recommended by the group, while any feasible alternatives must be discussed further.
- Conclusion: Following the previous step, the facilitator proposes the set of kept alternatives as the conclusion of the meeting. If the group must decide on only one alternative, it is still possible to go back to the step Multi-Criteria evaluation in order to refine the solution.
- Report: The facilitator generates a report of the meeting as a pdf file.

## 4 Results

Each of the three groups agreed on four criteria, as shown in Table 1. Each group identified four alternatives (not shown).

**Table 1.** Group sessions

| Group | Number of participants | Criteria selected | Number of alternatives identified |
|-------|------------------------|-------------------|-----------------------------------|
| 1 | 4 | Price<br>Operating System<br>Communication Autonomy & Battery Capacity<br>RAM | 4 |
| 2 | 4 | Price<br>Battery<br>Communication<br>Operating System | 4 |
| 3 | 6 | Price<br>Autonomy<br>RAM<br>Handling | 4 |

The survey results for all groups are summarized and discussed next.

### 4.1 Survey Results: Common vs. Private Criteria

The questionnaire contained five questions about whether the decision makers would feel comfortable using only common criteria. The participants' answered on a scale including 4 degrees plus one response for those who have no opinion: Completely agree, Rather agree, Rather not agree, Not at all agree, Without opinion.

**Question 1:** Do you think it is difficult for the group to find a set of shared criteria?

No participant chose "No opinion." A large majority agreed that it is difficult for a group to find shared criteria, as shown in Fig. 1.

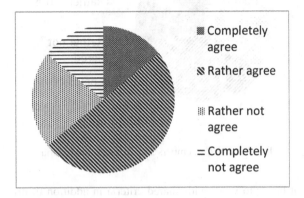

**Fig. 1.** Difficulty of finding shared criteria

**Question 2:** Do you think that group size makes it difficult for the group to find shared criteria?

Every participant reported an opinion. A large majority agreed that group size influences the group's ability to find shared criteria, as shown in Fig. 2.

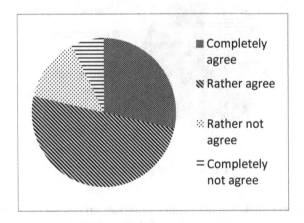

**Fig. 2.** Size of group influences ability to find shared criteria

**Question 3:** Do you think it should be mandatory for all group members to use the same criteria?

Again, every participant reported an opinion. A majority agreed that it should be mandatory for the group to work with a common set of criteria, as shown in Fig. 3.

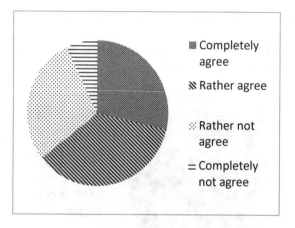

**Fig. 3.** Group members should use same criteria

**Question 4:** Is it better to work with shared criteria in addition to private criteria for individual decision makers?

Every participant reported an opinion. A large majority felt that using private criteria would help decision makers, as shown in Fig. 4.

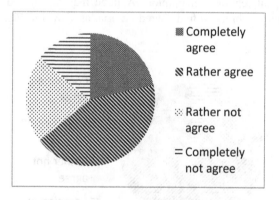

**Fig. 4.** Use of private criteria

**Question 5:** Do you think that the number of private criteria for each decision maker should be at least as great as the number of shared criteria?

Every participant offered an opinion. The results were balanced; half of the respondents supported equal numbers of private and shared criteria. Of the remainder, slightly more than half suggested that the number of private criteria should exceed the number of public criteria. The results are shown in Fig. 5.

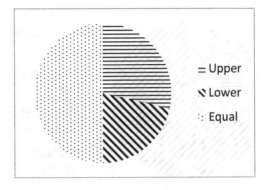

**Fig. 5.** Number of private criteria equal to number of shared criteria

## 4.2   Survey Results: Facilitation

The participants were asked two questions about the facilitation process. Their responses were on a scale including 4 degrees, plus one level for those who have no opinion: Completely agree, Rather agree, Rather not agree, Not at all agree, Without opinion.

**Question 6:** Do you think that GRUS should be used without a facilitator?

Every participant reported an opinion. A small majority agreed that the system could be used without a facilitator, as shown in Fig. 6.

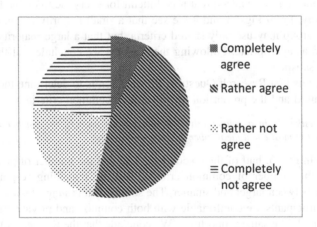

**Fig. 6.** Use of the system without facilitator

**Question 7:** Do you think that a decision process using the GRUS system is enough to support a group decision meeting?

No participant completely agreed, but a substantial majority of those with an opinion agreed that the system could be used with the work process it incorporates, as shown in Fig. 7.

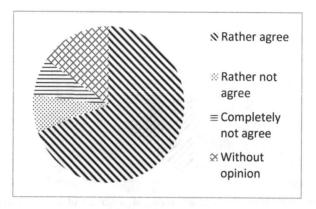

**Fig. 7.** Use of GRUS with no additional work process

## 5 Discussion

The Research Questions were assessed in light of the results obtained in the survey.

*Research Question 1: In the design of a collaborative decision process, participants benefit from the availability of both private and common criteria.*

Most participants find it difficult to identify common criteria (see Fig. 1) and agree that the size of the group influences its ability to find common criteria (see Fig. 2). Thus, the participants were aware that it is difficult for everyone to define the problem in the same way. From Figs. 3 and 4 we see that a small majority of the participants agree that the group may use only shared criteria, but that a large majority sees some private criteria as appropriate. Following these results we conclude that the Research Question 1 is satisfied.

Following this first Research Question, the question is to determine the number of criteria to be used and the proportion of private and common criteria.

*Research Question 2: In a collaborative decision making process, the number of private criteria should at most equal the number of common criteria.*

As shown in Fig. 5, half of the respondents feel that the number of private criteria should equal the number of common criteria. But the remaining respondents split almost equally between larger and smaller. The survey results suggest – see Figs. 3 and 4 – that the participants are comfortable with both common and private criteria, when they are in roughly the same proportions. We conclude that the Research Question 2 is weakly verified.

GDSS use normally involves a facilitator, who may be replaced by a computer system. The next Research Question aims to assess whether the participants feel that it is a better option.

*Research Question 3: GDSS use remains difficult without a human facilitator.*

Figures 6 and 7 show that the participants appreciate the contribution of a human facilitator, but believe that an automated system can help too. So we cannot draw any conclusions from our survey about the status of Research Question 3. We can only say that if an automated process is implemented to support the group, it could help, but that a human facilitator may also be helpful.

# 6 Concluding Remarks and Perspectives

Group decisions can be complex and may involve a large degree of conflict. Participants may feel dissatisfied because their wishes and views were not properly considered. They may not be motivated to participate because of an unwillingness, for strategic or privacy reasons, to reveal their assessment of the decision problem. Our view is that the use of both private and common criteria in Multi-Criteria Group Decisions can improve both participation and satisfaction.

Our aim is to study the use of private criteria in a group decision making process. This is only a preliminary study. It is obvious that the quality of the choices made by groups, as well as the range of alternatives and criteria that they generate, must still be studied in order to draw stronger conclusions about the potential contribution of private criteria to a group decision.

This study aimed to test the effects of using private and common criteria in group decisions and it is an exploratory work. This is the first step of a more global experiment including more participants. Clearly our experiments and surveys involved so few participants that no statistical significance can be attributed to our conclusions. In the future, we aim to involve more participants in order to deny or confirm our first results.

The results addressed certain factors that require careful consideration in the design of group decision processes and group decision support. One such factor is the impact of the homogeneity of the group. Cohesive groups can agree more easily, especially if there are dominant leaders, but the consequence is to limit creative solutions. Another concern that could be tested is the view that the use of GDSS reduces complexity, not only because of the larger numbers of group members, but also because the only way to find shared criteria is to look for the "lowest common denominator." Cultural effects could also influence the results, and it is our intention to test them by conducting other experiments in other countries.

Another contribution of this work could be to detail the role of the facilitator in supporting a group decision process: Which of the four presented tasks is in fact the most helpful (as judged by the participants and by the resulting decisions)? Knowledge of what is most important in facilitation would not only help human facilitators, it would be relevant to the design of automated facilitation and automated support.

**Acknowledgments.** The authors would like to thank the CIMI Excellence Laboratory, Toulouse, France, for inviting Marc Kilgour on a Scientific Expert position during the period May-June 2015 and for allowing a visiting period to Pascale Zaraté at the Wilfrid Laurier University, Waterloo, Canada, during the period December 2015 – February 2016.

# References

1. Gorry, G., Scott Morton, M.: A framework for management information systems. Sloan Manag. Rev. **13**(1), 50–70 (1971)
2. Smoliar, S., Sprague, R.: Communication and understanding for decision support. In: Proceedings of the International Conference IFIP TC8/WG8.3, Cork, Ireland, pp. 107–119 (2002)
3. Zaraté, P.: Tools for Collaborative Decision-Making. Wiley, Hoboken (2013)
4. Macharis, C., Brans, J.P., Maréchal, B.: The GDSS PROMETHEE procedure. J. Decis. Syst. **7**, 283–307 (1998)
5. Camilleri, G., Zaraté, P.: EasyMeeting: a group decision support system (Release 1). Rapport de recherche IRIT/RR–2014–10—FR (2014)
6. Yager, R.: On ordered weighted averaging aggregation operators in multicriteria decision making. IEEE Trans. Syst. Man Cybern. **18**, 183–190 (1988)
7. Choquet, G.: Theory of capacities. Ann. de l'Institut Fourier **5**, 131–295 (1953)
8. Schmidt, K., Bannon, L.: Taking CSCW seriously: supporting articulation work. Comput. Support. Coop. Work (CSCW) **1**(1), 7–40 (1992)
9. Sibertin-Blanc, C., Zaraté, P.: A flexible multi-criteria methodology for collective decision making preferences. Group Decis. Negot. J., to appear
10. Schwarz, R.: The Skilled Facilitator. Jossey-Bass Publishers, San Francisco (1994)
11. Ackermann, F., Eden, C.: Issues in computer and non-computer supported GDSSs. Decis. Support Syst. **12**, 381–390 (1994)
12. French, S.: Web-enabled strategic GDSS, e-democracy and Arrow's theorem: a Bayesian perspective. Decis. Support Syst. **43**(4), 1476–1484 (2007)
13. Limayem, L., DeSanctis, G.: Providing decisional guidance for multicriteria decision making in groups. Inf. Syst. Res. **11**(4), 386–401 (2000)
14. Nunamaker, J., Briggs, R.O., Mittleman, D., Vogel, D., Balthazard, B.: Lessons from a dozen years of group support systems research: a discussion of lab and field findings. J. Manag. Inf. Syst. **13**(3), 20–163 (1997)

# Plurality, Borda Count, or Anti-plurality: Regress Convergence Phenomenon in the Procedural Choice

Takahiro Suzuki$^{(\boxtimes)}$ and Masahide Horita

Department of International Studies, Graduate School of Frontier Sciences,
The University of Tokyo, Kashiwa, Japan
k147628@inter.k.u-tokyo.ac.jp, horita@k.u-tokyo.ac.jp

**Abstract.** We focus on voters' preference profiles where at least two of the three selected voting rules (e.g. plurality, Borda count, and anti-plurality) produce different outcomes—thus, the voting body needs a procedural choice. While this situation evokes an infinite regress argument for the choice of rules to choose rules to choose rules to…and so on, we introduce a new concept named regress convergence, where every voting rule in the menu ultimately gives the same outcome within the finite steps of regress. We study the mechanism of this phenomenon in a large consequential society having a triplet of scoring rules. The results show that, in the menu of plurality, Borda count, and anti-plurality, the probability that the regress convergence happens is 98.2% under the Impartial Culture assumption and 98.8% under the Impartial Anonymous Culture assumption.

**Keywords:** Scoring rules · Consequentialism · Procedural choice

## 1 Introduction

In modern democratic societies, the choice of voting rules has an important role. Saari [17] shows that a single profile of ten candidates could result in millions of different rankings for ten candidates simply depending on the choice of scoring rule. Even when there are three candidates, Saari shows that a preference profile could support up to seven different rankings[1] by changing the scoring rule. Nurmi [14] also argues the discrepancies within popular voting rules, the plurality, Borda count, max-min, and Copeland's method. Therefore, in addition to the choice of candidates, society must also choose their voting rules, which will also require to choose a voting rule for the choice of a voting rule. In this way, procedural choice falls into an infinite regress.

To overcome this problem, a growing number of studies have been carried out both from axiomatic and probabilistic viewpoints. Barbera and Jackson [2] and Koray [10] axiomatically studied the properties, called self-stability and self-selectivity, of a single voting rule for two and three or more alternatives, respectively. These concepts demand that a voting rule should choose itself among the other voting rules at hand in the

---

[1] If the alternatives are denoted $A$, $B$, and $C$, the seven rankings are $A \succ B \succ C$, $A \succ C \succ B$, $C \succ A \succ B$, $C \succ B \succ A$, $A \succ B \sim C$, $A \sim C \succ B$, and $C \succ A \sim B$ (Saari and Tataru 1999).

© Springer International Publishing AG 2017
D. Bajwa et al. (Eds.): GDN 2016, LNBIP 274, pp. 43–56, 2017.
DOI: 10.1007/978-3-319-52624-9_4

voting on the voting rules. Later on, the concepts are extended to apply for the set of voting rules. A set of voting rules is (first-level) stable if for any profile there is exactly one rule that chooses itself (Houy [9]), or if there is at least one self-selective voting rule (Diss and Merlin [4]). Using the methods developed by Saari and Tataru [18] and Merlin et al. [13], Diss and Merlin [4] estimate the likelihood that the set of plurality (P), Borda (B) and anti-plurality (A) is stable under the Impartial Culture (IC) assumption within a large society. Diss et al. [3] also determines corresponding probabilities under the Impartial Anonymous Culture (IAC) assumption. These estimations show that the set of $\{P, B, A\}$ is stable with a probability 84.49% under IC and 84.10% under IAC within a large society.

The objective of the present article is to propose a new way to analyze and solve the regress argument. Specifically, we formulate a phenomenon, named regress convergence, where the regress argument is supposed to naturally disappear within the finite steps of regress, and we show that this phenomenon occurs quite frequently in the choice of triplets of scoring rules. The regress convergence is a phenomenon where every voting rule in the agenda ultimately designates the same outcome.[2] Let us explain this with an example. Suppose a society of 14 individuals chooses one of three candidates $a, b, c$, and there is an ex ante agreement on the set of voting rules, $F = \{P, B, A\}$. When the preference profile on the set of candidates $X$ is given as $L^0_{1-10}$ : $abc$, and $L^0_{11-14}$ : $bca$ (individuals $1, 2, \ldots, 10$ prefer $a$ to $b$ and $b$ to $c$ and individuals $11, 12, 13, 14$ prefer $b$ to $c$ and $c$ to $a$), the three voting rules $P, B$, and $A$ yield $\{a\}, \{a\}$, and $\{b\}$, respectively. Suppose now that the same society votes on which rule in $F$ to use. If everyone is consequential (i.e., preferring those rules that yield better candidates for themselves) and is supposed to submit a linear ordering, it is natural to think that the first 10 individuals submit either "PBA" or "BPA," and the remaining four individuals submit "APB" or "ABP". Suppose that they submit the following: $L^1_{1-4}$ : PBA, $L^1_{5-10}$ : BPA, and $L^1_{11-14}$ : APB. For this profile $\left(L^1_1, L^1_2 \ldots, L^1_{14}\right)$, $P$ yields $\{B\}$ while $B$ and $A$ yield $\{P\}$ (See Fig. 1).

Note that each $P^2, B^2$, and $A^2$ (a rule to choose the rule) ultimately reach the same outcome $\{a\}$. This means that no matter which rule in $F^2$ is selected, the outcome is the same. Hence, further regress has no meaning for the determination of the ultimate outcome. In a large and consequential society, our result shows that the regress convergence phenomenon is not so rare. Indeed, under the menu of $\{P, B, A\}$, for example, the phenomenon occurs at more than 98% under either IC or IAC (Corollary 1).

The present article is organized as follows. Section 2 introduces the notation. Section 3 states basic results, and some of the results are expanded in the discussion found in Sect. 4. The conclusion is stated in Sect. 5, and all proofs are in the Appendix.

---

[2] Saari and Tataru [18] argue in their introduction that "Except in extreme cases such as where the voters are in total agreement, or where all procedures give a common outcome, it is debatable how to determine the 'true wishes' of the voters." Clearly, the intuition of regress convergence lies in the latter "extreme cases," though our results show that the phenomenon can occur relatively frequently in the choice of triplets of scoring rules.

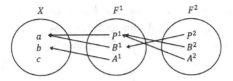

**Fig. 1.** Example of a regress convergence.[3] $F^1$ denotes the set of voting rules for the choice of candidates and $F^2$ denotes the set of voting rules for the choice of $F^1$.

## 2 Notation

Suppose a society $N = \{1, 2, \ldots, n\}(n \geq 3)$ makes a collective decision over the choice of $m$ alternatives without an agreement on the Social Choice Rule (SCR). Let $X = \{x_1, \ldots, x_m\}$ be the set of alternatives. Suppose also that they have in their mind $m$ possible SCRs, denoted by $f_1, f_2, \ldots, f_m$. We call this combination a menu of SCRs. For any nonempty set $A$, we denote by $\mathcal{L}(A)$ the set of all linear orderings over $A$.

Each individual $i \in N$ is supposed to have a linear preference $L_i^0 \in \mathcal{L}(X)$. The combination of $L_1^0, \ldots, L_n^0$ is called a level-0 preference profile. An SCR $f$ over a nonempty set $A$ is a correspondence $f : (\mathcal{L}(A))^n \twoheadrightarrow A$ such that $\phi \neq f(L) \subseteq A$ for all $L \in (\mathcal{L}(A))^n$. A scoring SCR $f$ for $m$ options is an SCR that assigns to each alternative $s_j(j = 1, 2, \ldots, m)$ points if it is ranked at the $j^{th}$ position in the preferences, where $1 = s_1 \geq s_2 \geq \cdots \geq s_m \geq 0$[4]. Then, $f(L)$ is defined as the set of options with the highest scores. We often express score assignments as $f : [s_1, s_2, \ldots, s_m]$. If $m = 3$, the plurality rule $f_P$ has the assignment $[1, 0, 0]$, the Borda count $f_B$ has the assignment $[1, 1/2, 0]$, and the anti-plurality rule $f_A$ has the assignment $[1, 1, 0]$. For any $m \in \mathbb{N}$, a $k$-approval voting $f_{E_k}$ is a scoring SCR such that $s_1 = s_2 = \cdots = s_k = 1$ and $s_{k+1} = s_{k+2} = \cdots = s_m = 0$.

The regress argument, i.e., the choice of SCRs for the choice of SCRs for the choice of...and so on, is supposed to be as follows (the terms in italics are formally defined later). It starts from the choice of $X$ using the set of SCRs $F^1$ (*level-1 issue*). If the society finds a *regress convergence*, then the regress stops. Otherwise, the society goes up to the choice of $F^1$ using the set of SCRs $F^2$ (*level-2 issue*). If the society finds a regress convergence, then the regress stops. Otherwise, the society makes the choice of $F^2$ using $F^3$ (*level-3 issue*), and so on there is regress convergence at some level. Note also that each individual is supposed to be *consequential* throughout the regress process.

---

[3] Note that no rule chooses itself in the figure. Therefore, the weak convergence does not logically imply the stability of the menu of SCRs in either Houy [9]'s or Diss and Merlin [4]'s sense. We can also say that the existence of a self-selective rule does not imply regress convergence (See the trivial deadlock described in Definition 5 and Fig. 2).

[4] Note that by definition we normalize the score assignment so that the top position gains one point and the worst position gains zero points.

**Definition 1 (Level[5]).** Let $X = F^0$. For any $k \in \mathbb{N} \cup \{0\}$, the level-$k$ issue is the choice from $F^{k-1}$ using $f_1, f_2, \ldots, f_m$. At this level, each SCR $f_j (j = 1, 2, \ldots, m)$ is called a level-$k$ SCR and is often denoted by $f_j^k$. We denote by $F^k = \{f_1^k, \ldots, f_m^k\}$ the set of level-$k$ SCRs.

**Definition 2 (Class $C \subseteq X$ of $f^k \in F^k$).** For any level-1 SCR $f \in \mathcal{F}^1$, its class at $L^0 \in (\mathcal{L}(X))^n$ is $f(L^0)(\subseteq X)$. For $k \geq 2$, the class of $g \in \mathcal{F}^k$ at $L^0 \in (\mathcal{L}(X))^n$, $L^1 \in (\mathcal{L}(F^1))^n$, $\ldots, L^{k-1} \in (\mathcal{L}(F^{k-1}))^n$, denoted $C_g[L^0, \ldots, L^{k-1}]$ or simply $C_g$, is the union of each class of $h \in g(L^{k-1})$ at $L^0, L^1, \ldots, L^{k-2}$.

**Definition 3 (Consequentialism[6]).** Let $k \in \mathbb{N}$ and $L^j \in (\mathcal{L}(F^j))^n (j = 0, 1, \ldots, k-1)$. $L^k \in (\mathcal{L}(F^k))^n$ is called a consequentially induced level-$k$ profile from $L^0, \ldots, L^{k-1}$ if for all $i \in N$, $x, y \in X$, $f, g \in F^k$, if $C_f = \{x\}$, $C_g = \{y\}$, and $xL_i^0 y$, then $fL_i^k g$. We denote by $\mathcal{L}^k[L^0, \ldots, L^k]$ the set of all such profiles.

When $L^0, L^1, \ldots, L^k$ is a sequence of profiles such that $L^j$ ($j = 1, \ldots, k$) is a consequentially induced level-$j$ profile from the preceding profiles, we simply say $L^0, \ldots, L^k$ as a consequential sequence of profiles. We are now ready to formally state how the regress argument could stop. The weak regress convergence (Definition 4) and the trivial deadlock (Definition 5) are thought to be a success and failure, respectively, in the procedural regress argument. In either case, further regress is thought to be meaningless.

**Definition 4 (Weak Regress Convergence).** Let $\{f_1, \ldots, f_m\}$ be the menu of SCRs. A level-0 preference profile $L^0 = (L_1^0, L_2^0, \ldots, L_n^0) \in (\mathcal{L}(X))^n$ weakly converges to $C \subseteq X$ if and only if a consequential sequence of profiles $L^0, L^1 \ldots, L^k$ exist such that $f_1^k, f_2^k, \ldots, f_m^k$ are all in the same class $C$ at $L^0, \ldots, L^k$.

**Definition 5 (Trivial Deadlock).** Let $\{f_1, \ldots, f_m\}$ be the menu of SCRs. A level-0 preference profile $L^0 \in (\mathcal{L}(X))^n$ is said to cause a trivial deadlock if and only if $f_1(L^0), f_2(L^0), \ldots, f_m(L^0)$ are mutually distinct singletons.[7]

**Example 1.** Suppose $m = 3$ and the menu of SCRs is $\{f_P, f_B, f_A\}$. If $f_P^1(L^0) = \{x_1\}$, $f_B^1(L^0) = \{x_2\}$, and $f_A^1(L^0) = \{x_3\}$ (as in Fig. 2), it is clear that for all $k \in \mathbb{N}$, a

---

[5] In this article, we suppose the society uses the fixed set of SCRs, $f_1, \ldots, f_m$ for any level. The distinction between $f_j^1$ and $f_j^2$ by the superscripts is made based on the supposed agenda.

[6] If we identify $f \in F^k$ with its class $C \subseteq X$, the consequentialism assumption is a way to introduce one's preference on sets of alternatives. This is often called preference extension (Barbera et al. [1]). When seen in this way, our consequentialism assumption is the same as the Extension Rule. It is a natural requirement of most reasonable systems of preference extension to satisfy the Extension Rule (See also Endriss [6]).

[7] In the present article, we adjust the number of alternatives and that of admissible SCRs. However, this is not essential. If the society has $m'(\neq m)$ alternatives and $m$ scoring SCRs, we can modify the definition of trivial deadlock to be the case where every level-2 scoring SCR chooses distinct singletons and $f_1^2(L^1) \cup \ldots \cup f_m^2(L^1) = F^1$ for all $L^1 \in \mathcal{L}^1[L^0]$. Then, our proofs for Lemma 1 and Proposition 1 also hold (though the specific value of $p_D$ depends on $m'$).

consequentially induced $L^k$ is unique and $f_P^{k+1}(L^k) = \{f_P^k\}$, $f_B^{k+1}(L^k) = \{f_B^k\}$, and $f_A^{k+1} = \{f_A^k\}$. Therefore, no matter how high of a level we see, the structure does not change at all, which makes the regress argument meaningless in a negative sense (Fig. 2).

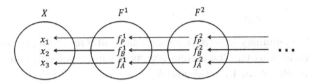

**Fig. 2.** A graph image of trivial deadlock

Finally, we discuss the asymptotic probabilities as $n \to \infty$. Among the probabilistic studies of voting events, there are two major assumptions on the distribution of preferences. One is called the Impartial Culture (IC). It assumes that each voter independently chooses, with equal likelihood, one of the linear orderings over $X$. Therefore, each profile $L^0 \in (\mathcal{L}(X))^n$ occurs with probability $1/(m!)^n$. The other assumption is called the Impartial Anonymous Culture (IAC). This assumes that every voting situation, a combination of the numbers of individuals who each have a specific linear ordering, occurs with equal likelihood. Hence, each $(n_1, \ldots, n_{m!})$, where $n_j$ represents the number of individuals who have the $j^{th}$ linear ordering, occurs with the probability $1/{_{n+m!-1}C_n}$. While the probability of certain events, such as the Condorcet winner exists, can differ according to which assumption is used, there are two common properties when $n \to \infty$. The first common property is that the probability that a specific scoring rule yields a tied outcome is negligible as $n \to \infty$[8] (Marchant [12]; Pritchard and Wilson [15]; Pritchard and Wilson [16]; Diss and Merlin [4]). Let us denote by $p_l$ the probability that at least one level-1 SCR in $F^1$ yields a tied outcome. Because IC and IAC both say that $p_l \to 0$ as $n \to \infty$, we can focus only on the cases where each level-1 SCR yields a singleton outcome. The second common property is that the probability that exactly $\alpha \in [0, 1]$ of the whole individuals prefer $x$ to $y$ for some $x, y \in X$, where $\alpha$ is a fixed constant, is negligible as $n \to \infty$. We show this in the appendix (Lemma 3).[9] We denote by $q_\alpha$ the probability that exactly $\alpha$ of the whole individuals prefer $x$ to $y$ for some $x, y \in X$.

---

[8] For a relatively small $n$, the probabilities of tied outcomes by famous scoring rules such as plurality and Borda count are studied by Gillet [7,8] and Marchant [12].

[9] Note that these two properties (and thus, our Proposition 1) are also satisfied under IANC (impartial anonymous and neutral culture) model introduced by Eğecioğlu [5]. This is because each ANEC (anonymous and neutral equivalence class) has at most $m!$ different AECs. Hence, the ratio of those ANECs including profiles such that ties happen or $\#\{i \in N | xL_i y\} = n\alpha$ to the whole ANECs is at most $(m!)^2$ times that of IAC, i.e. the ratio of AECs (anonymous equivalence class) causing ties or $\#\{i \in N | xL_i y\} = n\alpha$ to the whole AECs. With our Lemma 3, we can confirm that the two asymptotic properties still hold under IANC. We thank an anonymous referee for encouraging us to consider about IANC also.

In sum, either under IC or IAC, the probabilities $p_l$ and $q_\alpha$ (for several $\alpha$ s) are both negligible as $n \to \infty$. Based on this property, we next evaluate $p_{WC}$, the probability that those level-0 profiles occur that weakly converge, and $p_D$, the probability that those occur that cause a trivial deadlock.

# 3 Results

Beginning with preliminary lemmas, we show our central result Proposition 1. It shows under several conditions that the regress argument can be solved (with weak convergence) unless it falls in the trivial deadlock.

**Lemma 1.** Let $n \geq m$ and $F = \{g_1, g_2, \ldots, g_p, h_1, h_2, \ldots, h_q\}$ $(p \geq q \geq 0)$ be the menu of scoring SCRs, where $m = p + q \geq 3$. For given consequential sequence of profiles $L^0, \ldots, L^{k-1}$ and alternatives $x, y \in X$, suppose the following holds:

$$C_{g_j^k}[L^0, L^1, \ldots, L^{k-1}] = \{x\} \text{ for all } j = 1, 2, \ldots, p \tag{1}$$

$$C_{h_j^k}[L^0, L^1, \ldots, L^{k-1}] = \{y\} \text{ for all } j = 1, 2, \ldots, q. \tag{2}$$

If $\#\{i \in N | x L_i^0 y\} > \#\{i \in N | y L_i^0 x\}$, then $L^0$ weakly converges to $\{x\}$.

**Lemma 2.** Let $n \geq m$, $m = 3$ or $4$, and $x, y \in X$ such that $\#\{i \in N | x L_i^0 y\} \neq n/2$. If the menu of SCRs is $F = \{f_{E_1}, f_{E_2}, \ldots, f_{E_{m-1}}, f_B\}$ and the class of each level-$k$ SCR is either $\{x\}$ or $\{y\}$ at given $L^0, \ldots, L^{k-1}$, then $L^0$ weakly converges.

**Proposition 1.** Suppose $m = 3$ and $n$ is large ($n \to \infty$). Denote the three scoring SCRs as $f_i : [1, s_i, 0]$ $(i = 1, 2, 3)$, where $1 \geq s_1 > s_2 > s_3 \geq 0$. Either under IC or IAC, we have $p_{WC} \approx 1 - p_D$ whenever the following holds:

$$s_3 \geq 1/2 \text{ or } \left[ s_3 < 1/2 \text{ and } s_2 \leq \frac{1 + s_3}{2 - s_3} \right]. \tag{3}$$

It is worth noting that if the menu of SCRs contains $\{f_B, f_A\}$ or $\{f_P, f_B\}$, then the condition 0 automatically holds irrespective of the last one. Therefore, once a large consequential society admits the menu $\{f_P, f_B, f_A\}$, for example, Proposition 1 shows there are approximately only two possibilities: they face a trivial deadlock, or they are endowed with the ability to realize the weak convergence. The probability $p_D$ under IC and IAC is determined by Diss and Merlin [4] and Diss et al. [3]. Based on their probability calculation, we have the following.

**Corollary 1.** Let $n \to \infty$ and $m = 3$, where the menu of SCRs is given by $\{f_P, f_B, f_A\}$. Under IC, the regress weakly converges with a probability of 98.2%. Under IAC, the regress weakly converges with the probability of 98.8%.

# 4 Discussion

## 4.1 Uniqueness of the Convergent Outcome

$L^0 \in \mathcal{L}(X)$ is said to weakly converge if at least one consequential sequence of (sub-sequent) profiles $L^1, L^2, \ldots$ exists[10] that adjust the rules' ultimate judgments at a certain level. The existence of such $L^1, L^2, \ldots$ guarantees that we can stop the apparent infinite regress arguments through finite steps of regress. One might, however, be concerned that the same $L^0$ might weakly converge to a distinct $C$ and $C'$ by the choice of the sequence. Now, we show that the set of $\{f_P, f_B, f_A\}$ guarantees the uniqueness of the convergent outcome with a slightly stronger assumption on the meta-preferences.

**Definition 5 (Strong Convergence).** Let $\{f_1, \ldots, f_m\}$ be the menu of SCRs. A level-$0$ preference profile $L^0 \in (\mathcal{L}(X))^n$ strongly converges to $C \subseteq X$ if and only if it weakly converges to $C$, and if it does not weakly converge to any other $C' \subseteq X$. (Note that strong convergence implies weak convergence but not vice versa.)

*Expected Utility assumption (EU).* For given $L^0 \in (\mathcal{L}(X))^n$ and their utility representations $u_i^0 : X \to \mathbb{R}(i \in N)$, i.e. $u_i^0(x) \geq u_i^0(y) \Leftrightarrow xL_i^0 y$, the subsequent sequence of profiles $L^1, L^2, \ldots$ satisfies EU if they satisfy the following:

(1) Each $i \in N$ has utility function $u_i^k$ over $F^k$ such that $\forall f, g \in F^k (k \in \mathbb{N})$,

$$u_i^k(f) \geq u_i^k(g) \Leftrightarrow \frac{\sum_{x \in C_f} u_i^0(x)}{|C_f|} \geq \frac{\sum_{y \in C_g} u_i^0(y)}{|C_g|} \tag{4}$$

(2) Let $R_i^k$ be a weak ordering represented by $u_i^k$. $L_i^k$ is compatible with $R_i^k$ (i.e. $L_i^k$ is obtained by breaking the indifferences in $R_i^k$).

(Note that EU logically implies consequentialism (Definition 3), but not vice versa.)

Under EU, we modify Definition 4 by substituting "a consequential sequence of profiles $L^0, L^1, \ldots$" with "a sequence of profiles $L^0, L^1, \ldots$ satisfying EU".

**Proposition 2.** Assume $m = 3$, $n \to \infty$, EU, and either IC or IAC. Then, for the menu of SCRs $F = \{f_P, f_B, f_A\}$, we have $p_{SC} \approx 1 - p_D$, where $p_{SC}$ is the probability that $L^0$ occurs that strongly converges.

Based on Proposition 2, a large consequential society holding $\{f_P, f_B, f_A\}$ as the menu of SCRs has only two possibilities: either the society faces a trivial deadlock

---

[10] Technically speaking, we can find the similar use of a compatible linear ordering in Koray [10] and Koray and Slinko [11] . They define a Social Choice Function (SCF) $f$ as self-selective at $L^0$ relative to the menu of SCFs $F^1$ if and only if there is a consequentially induced $L^1 \in (\mathcal{L}(F^1))^n$ such that $f^2(L^1) = f^1$. As Koray and Slinko stated (if we impose that the rule chooses itself for all compatible linear orderings), "it leads to a vacuous concept". The same applies to regress convergence.

(with at most 1.8% under IC and 1.2% under IAC) or they can know the possible regress convergence without implementing the regress arguments.

## 4.2    Tie-Breaking by the Scoring Rules

Finally, we deal with the choice of Social Choice Function (SCF) (i.e., not a correspondence but a function). Suppose we provide SCRs with neutral tie-breaking systems. Especially, for any SCR $f_Y$, we denote by $f_{Y^*}$ the SCF that breaks ties in favor of $i_Y \in N$, named the tie breaker of $f_Y$. Note that different SCRs are allowed to have different tie breakers (for example, the plurality tie breaker $i_P = 1$ and the Borda count tie breaker $i_B = 2$). Then, Proposition 2 can be revised for a relatively small $n$ (it is straightforward to revise the proofs of Lemma 2 and Proposition 2, so we omit the proof).

**Proposition 3.** Let us assume $n$ is odd ($n \geq 3$), $m = 3$, and the menu of SCFs is either $\{f_{P^*}, f_{X^*}, f_{A^*}\}$, where $f_X$ is either the Borda count, Black's rule, Copeland's method, or the Hare system.[11] Then, any level-0 profile $L^0$ either causes trivial deadlock or strongly converges.

## 5    Conclusion

We analyzed the regress arguments for procedural choice in a large ($n \to \infty$) consequential society. Once the society admits the menu of SCRs (plurality, Borda count, and anti-plurality), the probability that at least two of them give different outcomes is about 46.5% under IC (from Table 7 of Diss and Merlin [4]). While this fact emphasizes the importance of procedural choice, Proposition 1 says that at more than 98% (either under IC or IAC), we can derive a weak convergence. Furthermore, our Proposition 2 and Proposition 3 show even further that there are ways to uniquely determine the possible convergent outcome.

A different interpretation of our results can be obtained when compared with Suzuki and Horita [19], who argue the difficulties of ranking meta-procedures with a menu of all the possible SCFs. On the other hand, the present paper shows that the difficulty of procedural choice quite frequently disappears when the society considers a relatively small menu of voting rules, such as plurality, Borda count, and anti-plurality. It can be an interesting future topic to determine the tradeoff between the size of the menu and the possibility of resolving the regress problem.[12]

**Acknowledgements.** This work was supported by *JSPS KAKENHI, grant number 15J07352.*

---

[11] We assume the Hare system drops exactly one alternative with the least plurality score in each round (if two or more get the least score, it selects and drops one of them in neutral way).

[12] As a step further in this direction, it is also shown that the asymptotic property of $p_{WC} \approx 1 - p_D$ holds in a large consequential society having $\{f_P, f_{E_2}, f_A, f_B\}(m = 4)$. The sketch of this calculation is found in the appendix.

# A Appendix

***Extra Notation***: If $f^k \in F^k$ is obvious in the context, we denote by $s(f^{k-1} : L^{k-1})$ the score of $f^{k-1} \in F^{k-1}$ at $L^{k-1}$ evaluated by $f^k$. If $L^{k-1}$ is also obvious, we write as $s(f^{k-1})$.

**Lemma 3.** For any $x, y \in X$ and $\alpha \in [0, 1]$, and either under IC or IAC assumption, $P(\alpha)$, the probability that exactly $n\alpha$ individuals prefer $x$ to $y$, converges to zero as $n \to \infty$.

**Proof of Lemma 3.** We can suppose $n\alpha \in \mathbb{N}$. [Under IC] We have the following:

$$P(\alpha) = \binom{n}{n\alpha} \left(\frac{1}{2}\right)^{n\alpha} \left(\frac{1}{2}\right)^{n(1-\alpha)} = \binom{n}{n\alpha} \left(\frac{1}{2}\right)^n.$$

Because the proof is similar, we show only for even $n$. Let $n = 2k(k \in \mathbb{N})$. Then,

$$P(\alpha) \leq P\left(\frac{1}{2}\right) = \binom{2k}{k} \left(\frac{1}{2}\right)^{2k} = \frac{(2k)!}{k!k!} \left(\frac{1}{2}\right)^{2k}.$$

Using Stirling's approximation, we can evaluate the RHS as

$$\lim_{\substack{k \to \infty \\ (\Leftrightarrow n \to \infty)}} \frac{(2k)!}{k!k!} \left(\frac{1}{2}\right)^{2k} = \lim_{k \to \infty} \frac{\sqrt{2\pi \cdot 2k}\left(\frac{2k}{e}\right)^{2k}}{\left(\sqrt{2\pi k}\left(\frac{k}{e}\right)^k\right)^2} \left(\frac{1}{2}\right)^{2k} = \lim_{k \to \infty} \frac{1}{\sqrt{\pi n}} = 0.$$

[Under IAC] Let $a = \#\{i \in N | xL_i^0 y\} = n\alpha$ and $b = n - a$. The probability is described as:

$$P(\alpha) = \binom{a + \frac{m!}{2} - 1}{a} \cdot \binom{b + \frac{m!}{2} - 1}{b} \bigg/ \binom{a + b + m! - 1}{a + b}.$$

With a simple calculation, this is shown to converge to zero as $n = a + b \to \infty$. ∎

**Proof of Lemma 1.** Assume that $F = \{g_1, \ldots, g_p, h_1, \ldots, h_q\}$ and $L^0, L^1, \ldots, L^{k-1}$ satisfy the given condition. Let $A = \{1, 2, \ldots, a\} = \{i \in N | xL_i^0 y\}$. If $q = 0$, the lemma is obvious. So, we assume $p \geq q > 0$. It follows that $0 < |A| = a < n$ (if $a = 0$, e.g., no level-1 SCR chooses $\{x\}$, which contradicts with $p > 0$). Since $n \geq m$, we have $a \geq (n/2) \geq (m/2) \geq q$. Let $L^k$ be a profile on $F^k$ defined as follows:

$$L_i^k : g_1^k, g_2^k, \ldots, g_p^k, h_1^k, h_2^k, \ldots, h_{i-1}^k, h_{i+1}^k, \ldots, h_q^k, h_i^k \text{ for all } 1 \leq i \leq q$$

$$L_i^k : g_1^k, g_2^k, \ldots, g_p^k, h_1^k, h_2^k, \ldots, h_q^k \text{ for all } q + 1 \leq i \leq a$$

$$L_i^k : h_1^k, h_2^k, \ldots, h_q^k, g_1^k, g_2^k, \ldots, g_p^k \text{ for all } i \in N \backslash A.$$

In words, this is a level-$k$ profile where everyone (except the first $q$ individuals) orders $\{g_1^k, \ldots, g_p^k\}$ and $\{h_1^k, \ldots, h_q^k\}$ lexicographically. Clearly, we have $L^k \in \mathcal{L}^k[L^0, \ldots, L^{k-1}]$. Take any $f^{k+1} : [1 = s_1, s_2, \ldots, s_m = 0] \in F^{k+1}$ and consider the scores evaluated by this $f^{k+1}$. Note that $h_1^k$ has the largest score among $h_1^k, \ldots, h_q^k$. We have

$$
\begin{aligned}
s(g_1^k) - s(h_1^k) &= \{a + (n-a)s_{q+1}\} - \{n - a + (a-1)s_{p+1}\} \\
&\geq 2a - n + (n-a)s_{q+1} - (a-1)s_{q+1} (\because p \geq q \Rightarrow s_{q+1} \geq s_{p+1}) \\
&= (2a-n)(1 - s_{q+1}) + s_{q+1} > 0 (\because 2a > n \text{ and } 0 \leq s_{q+1} \leq 1).
\end{aligned}
$$

Since this holds for any $f^{k+1} \in F^{k+1}$, we obtain the weak convergence to $\{x\}$ (via $L^k$). ∎

**Proof of Lemma 2.** Let $A = \{1, 2, \ldots, a\} = \{i \in N | xL_i^0 y\}$, $G := \{g | C_g = \{x\}\} = \{g_1^k, \ldots, g_p^k\}$ $(p = |G|)$ and $H := \{h | C_h = \{y\}\} = \{h_1^k, \ldots, h_q^k\}$ $(q = |H|)$. With Lemma 1, we have only to consider $0 < a < n - a$ and $p > q > 0$, i.e. $(p, q) = (2, 1)$ if $m = 3$ or $(p, q) = (3, 1)$ if $m = 4$. Since the proof is similar, we show for the latter, $m = 4$. We can check that for all $L^k \in \mathcal{L}^k[L^0, \ldots, L^{k-1}], f_{E_1}(L^k) \subseteq H$ and the scores (at $L^k$) satisfy

$$
S := s_B(g_1^k) + s_B(g_2^k) + s_B(g_3^k) = a(s_1 + s_2 + s_3) + (n - a)(s_2 + s_3 + s_4) = n + a.
$$

Let $p_1, \ldots, p_6$ be preferences over $G$ such that $p_1 : g_1^k g_2^k g_3^k$, $p_2 : g_3^k g_2^k g_1^k$, $p_3 : g_3^k g_1^k g_2^k$, $p_4 : g_2^k g_1^k g_3^k$, $p_5 : g_1^k g_3^k g_2^k$, and $p_6 : g_2^k g_3^k g_1^k$. We construct $L^k \in \mathcal{L}^k[L^0, \ldots, L^{k-1}]$ as follows: if $i \equiv j \pmod 6$ then $L_i^k|_G = p_j$ $(j = 1, 2, \ldots, 6)$, and $g_\mu^k L_i^k h_1^k (\mu = 1, 2, 3) \Leftrightarrow i \leq a$. Because of the symmetry, we obtain that $s_B(g_j^k : L^k) - S/3 \in \{-1/3, 0, 1/3\}$ $(j = 1, 2, 3)$. Hence,

$$
D(L^k) := s_B(h_1^k : L^k) - \max\{s_B(g_1^k : L^k), s_B(g_2^k : L^k), s_B(g_3^k : L^k)\} \geq \frac{2}{3}(n - 2a) - \frac{1}{3}.
$$

Since $n - 2a \geq 1$, we have $D(L^k) > 0$. (1) The case of $n - 2a \geq 2$. Then, we have $D(L^k) \geq 1$. Suppose $\{g, h_1^k\} \in f_{E_{j'}}(L^k)$ for some $g \in G$ and $j' = 2, 3$. Let $j$ be the smallest such $j'$. Since $s_j(h_1^k) = n - a < n$, there exists $i_g \in N$ whose $L_{i_g}^k$ assign zero point to $g$ and one point to $g' \in G \setminus \{g\}$. Let $L'^k$ be a profile where $i_g$ swaps $g$ and $g'$. Then, we have $s_j(g : L'^k) > s_j(g : L^k) = s_j(h_1^k : L^k) > s_j(h_1^k : L'^k)$. Therefore, $f_{E_j}(L'^k) = \{g\} \subseteq G$. Since the change in Borda score of $g_1^k, g_2^k, g_3^k$ is at most $2/3$, we still have $D(L'^k) \geq 1 - (2/3) > 0$. (2) The case of $n - 2a = 1$. Since $n$ is odd, we can write $n = 6\mu + v$, where $\mu \in \mathbb{N} \cup \{0\}$ and $v = 1, 3, 5$. Note that the swap of $L_i^k|_G$ and $L_j^k|_G$ for any $i, j \in N$ does not at all affect $s_1(\cdot)$ and $s_B(\cdot)$. If $n = 6\mu + 1$ ($\mu \geq 1$ since $n \geq m = 4$), let $\left(\mathcal{L}^{(1)}\right)^n \in \mathcal{L}^k\left[L^0,\ldots,L^{k-1}\right]$ be defined as: $1 \leq i \leq \mu \Rightarrow L^{(1)^k}_i : p_3$, $\mu+1 \leq i \leq 2\mu \Rightarrow L^{(1)^k}_i : p_4$, $2\mu+1 \leq i \leq 3\mu \Rightarrow L^{(1)^k}_i : p_5$, $3\mu+1 \leq i \leq 4\mu \Rightarrow L^{(1)^k}_i : p_1$, $4\mu+1 \leq i \leq 5\mu \Rightarrow L^{(1)^k}_i : p_2$, $5\mu+1 \leq i \leq 6\mu \Rightarrow L^{(1)^k}_i : p_6$, and $i = 6\mu+1 \Rightarrow L^{(1)^k}_i : p_1$. Then, we have $s_3\left(g_1^k : L^{(1)^k}\right) \geq s_2\left(g_1^k : L^{(1)^k}\right) = 3\mu+2 > 3\mu+1 = s_2\left(h_1^k : L^{(1)^k}\right) = s_3\left(h_1^k : L^{(1)^k}\right)$. Hence, it follows that $f_{E_2}^{k+1}\left(L^{(1)^k}\right) \subseteq G$ and $f_{E_3}^{k+1}\left(L^{(1)^k}\right) \subseteq G$. For the other cases of $n = 6\mu+3$ and $n = 6\mu+5$, the following $L^{(2)^k}$ ($g_3^k$ wins) and $L^{(3)^k}$ ($g_1^k$ wins), respectively, gives the corresponding inequalities. $L^{(2)^k}$ is defined as: $1 \leq i \leq \mu \Rightarrow p_4$, $\mu+1 \leq i \leq 2\mu \Rightarrow p_5$, $2\mu+1 \leq i \leq 3\mu \Rightarrow p_6$, $i = 3\mu+1 \Rightarrow p_1$, $3\mu+2 \leq i \leq 4\mu+1 \Rightarrow p_1$, $4\mu+2 \leq i \leq 5\mu+1 \Rightarrow p_2$, $5\mu+2 \leq i \leq 6\mu+1 \Rightarrow p_3$, $i = 6\mu+2 \Rightarrow p_2$, and $i = 6\mu+3 \Rightarrow p_3$. $L^{(3)^k}$ is defined as follows: $1 \leq i \leq \mu \Rightarrow p_2$, $\mu+1 \leq i \leq 2\mu \Rightarrow p_3$, $2\mu+1 \leq i \leq 3\mu \Rightarrow p_4$, $i = 3\mu+1 \Rightarrow p_3$, $i = 3\mu+2 \Rightarrow p_4$, $3\mu+3 \leq i \leq 4\mu+2 \Rightarrow p_1$, $4\mu+3 \leq i \leq 5\mu+2 \Rightarrow p_5$, $5\mu+3 \leq i \leq 6\mu+2 \Rightarrow p_6$, $i = 6\mu+3 \Rightarrow p_1$, and $i = 6\mu+4 \Rightarrow p_2$, and $i = 6\mu+5 \Rightarrow p_5$.

In either case above, at least 2 level-$(k+1)$ SCRs has class $\{x\}$ and the other two have either $\{x\}$ or $\{y\}$. So, we can apply Lemma 1 to get the weak convergence (if $m = 3$, with Lemma 4 and the technique shown above, we can verify $L^k \in \mathcal{L}^k\left[L^0,\ldots,L^{k-1}\right]$ such that $f_{E_1}^{k+1}(L^k) = f_B^{k+1}(L^k) = \{h_1^k\}$ and $f_{E_2}^{k+1}(L^k)$ is either $\{g_1^k\}$ or $\{h_1^k\}$).

**Lemma 4.** Let $m = 3$ and $F^k = \{g_1^k, g_2^k, g_3^k\}$, where $g_j^k : [1, s_j, 0]$. Assume $C_{g_1^k} = C_{g_2^k} = \{x\}$ and $C_{g_3^k} = \{y\}$. Then, there exists $L^k \in \mathcal{L}^k\left[L^0, L^1, \ldots, L^{k-1}\right]$ such that $\left|s_j(g_1^k : L^k) - s_j(g_2^k : L^k)\right| \leq 1$ for all $j = 1, 2, 3$, where $s_j()$ denotes the score evaluated by $g_j^{k+1}$.

**Proof of Lemma 4.** Let $A = \{1, 2, \ldots, a\} = \{i \in N | x L_i^0 y\}$. We assume that both $a$ and $n - a$ are odd. The cases where at least one of them is even can be similarly (and more simply) proven. Note that the fact that $C_{g_1^k} = \{x\}$ and $C_{g_3^k} = \{y\}$ guarantees $a > 0$ and $n - a > 0$. Let $L^k \in \left(\mathcal{L}(F^k)\right)^n$ be such that $g_1^k L_i^k g_2^k L_i^k g_3^k$ for all $i : 1 \leq i \leq \frac{a}{2} + \frac{1}{2}$, $g_2^k L_i^k g_1^k L_i^k g_3^k$ for all $i : 1 \leq i \leq \frac{a}{2} - \frac{1}{2}$, $g_3^k L_i^k g_2^k L_i^k g_1^k$ for all $i : 1 \leq i \leq \frac{n-a}{2} - \frac{1}{2}$, and $g_3^k L_i^k g_1^k L_i^k g_2^k$ for all $i : 1 \leq i \leq \frac{n-a}{2} + \frac{1}{2}$. Clearly, $L^k \in \mathcal{L}^k\left[L^0, L^1, \ldots, L^{k-1}\right]$. We have also that

$$\left|s(g_1^k) - s(g_2^k)\right| = |(1 - s) - s| = |1 - 2s|.$$

The assumption of $0 \leq s \leq 1$ indicates that this absolute value is at most one. ∎

**Proof of Proposition 1.** Take any distinct $x, y \in X$. Let $A = \{i \in N | x L_i y\}$ and $\alpha := |A|/n$. The only nontrivial case is such that $g_1^1(L^0) = g_2^1(L^0) = \{x\}$ and $g_3^1(L^0) = \{y\}$, where $F^1 = \{g_1^1, g_2^1, g_3^1\}$. Due to Lemma 1, we need only consider $\alpha < 1/2$. Take any $f : [1, s, 0] \in F^2$. With Lemma 4, we have the following:

$$s\big(g_3^1 : L^1\big) = n - |A|, \quad \max_{L^1 \in \mathcal{L}[L^0]} s\big(g_1^1 : L^1\big) = |A| + s(n - |A|)$$

$$\min_{L^1 \in \mathcal{L}[L^0]} \max\big\{\big\{s\big(g_1^1 : L^1\big), s\big(g_2^2 : L^1\big)\big\}\big\} \le \frac{1}{2}\big\{|A|(1+s) + (n - |A|)s\big\} + \frac{1}{2}.$$

Therefore, $f$ can choose $\{g_1^1\}$ (or $\{g_2^1\}$) if and only if

$$|A| + s(n - |A|) > n - |A| \Leftrightarrow s > \frac{n - 2|A|}{n - |A|} = \frac{1 - 2\alpha}{1 - \alpha} = \varphi(\alpha).$$

Also, $f$ can choose $\{g_3^1\}$ if

$$\frac{1}{2}\big\{|A|(1+s) + (n - |A|)s\big\} + \frac{1}{2} < n - |A|$$

$$\Leftrightarrow s < 2 - \frac{3|A|}{n} - \frac{1}{n} = 2 - 3\alpha - \frac{1}{n} (\to 2 - 3\alpha = \psi(\alpha) \text{ as } n \to \infty).$$

If $\alpha < 1/3$, we have $\psi(\alpha) > 1$. Thus, any scoring SCR $f : [1, s, 0]$ can choose $\{g_3^1\}$. If $1/3 < \alpha < 1/2$, we have three cases. (note that events such as $\alpha = 1/3$ <u>or</u> $\psi(\alpha) - 1/n < s < \psi(\alpha)$ can be negligible because of Lemma 3). <u>(1) The case of</u> $s_3 \ge \varphi(1/3) = 1/2$. Then, each $f_1^2, f_2^2, f_3^3$ can exclude $g_3^1$ for any $\alpha \in (1/3, 1/2)$. <u>(2) The case of</u> $s_3 < \varphi(1/3)$ <u>and</u> $s_2 \le \psi(\varphi^{-1}(s_3))$. Note that the event $\alpha = \varphi^{-1}(s_3)$ is negligible because of Lemma 3. If $1/3 < \alpha < \varphi^{-1}(s_3)$, we have $\psi(\alpha) > s_2$, which implies that $L^1 \in \mathcal{L}^1[L^0]$ exists such that $f_2^2(L^1) = f_3^2(L^1) = \{g_3^1\}$ and $f_1^2(L^0)$ is either $\{g_1^1\}$ or $\{g_3^1\}$. In either case, with Lemma 1, $L^0$ is shown to weakly converge to $\{y\}$. If $\varphi^{-1}(s_3) < \alpha < 1/2$, $L^1 \in \mathcal{L}[L^0]$ exists such that $f_1^2(L^1) = f_2^2(L^1) = f_3^2(L^1) = \{g_1^1\}$. <u>(3) The case of</u> $s_3 < \varphi(1/3)$ <u>and</u> $s_2 > \psi(\varphi^{-1}(s_3))$. In this case, an interval of $\alpha$ (with a positive Lebesgue measure) exists where $f_1^1$ and $f_2^1$ necessarily choose $\{g_1^1\}$ or $\{g_2^1\}$ and $f_3^2$ necessarily chooses $\{g_3^1\}$. If $\alpha$ is in this interval, we cannot solve the regress, because inductively we can show for all $k \ge 3$ that $f_1^k(L^{k-1})$ and $f_2^k(L^{k-1})$ are either $\{f_1^{k-1}\}$ or $\{f_2^{k-1}\}$ and $f_3^k(L^{k-1}) = \{f_3^{k-1}\}$. ∎

**Proof of Corollary 1.** Under IC, trivial deadlock corresponds with cases 1, 2, 9, 10, 11, and 27 in Diss and Merlin [4]. Their Table 7 (p. 302) shows that each probability is 0.00299346. Therefore, $p_D = 0.00299346 \times 6 \fallingdotseq 1.8\%$. Under IAC, trivial deadlock corresponds with the cases 1, 2, 9, 10, 11, and 27 in Diss et al. [3]. Their Table 9 (p. 62) shows that each probability is 1/504. Therefore, $p_D = (1/504) \times 6 \fallingdotseq 1.2\%$. ∎

**Proof of Proposition 2.** The only nontrivial case is $f_1^1(L^0) = f_2^1(L^0) = \{x\}$ and $f_3^1(L^0) = \{y\}$, where $F^1 = \{f_1^1, f_2^1, f_3^1\}$ for distinct $x, y \in X$. Let $A = \{i \in N | xL_i^0 y\} = \{1, 2, \ldots, a\}$. We show that $L^0$ strongly converges unless $\alpha$ takes several specific values. The case of $\alpha > 2/3$ or $\alpha < 1/3$ is straightforward. Because the proof is similar, we show for $1/3 < \alpha < 1/2$. To prove the uniqueness of convergence to $\{y\}$, we inductively show that for any level $k \ge 2, f^k \in F^k$ exists whose class is $\{y\}$. For $k = 2$, it follows that $f_P^2(L^1) = \{f_3^1\}$. Assume that the statement holds until $k-1 (\geq 2)$ and $C_{g_1^{k-1}} = \{y\}$. For the other two rules $g_2^k$ and $g_3^k$, the class is either $\{x\}, \{x,y\}$, or $\{y\}$. Because $g_2^{k-1}$ and $g_3^{k-1}$ are symmetric, there are six possible cases on the combination of $\left(C_{g_1^{k-1}}, C_{g_2^{k-1}}, C_{g_3^{k-1}}\right)$: Case 1: $(\{y\},\{x\},\{x\})$, Case 2: $(\{y\},\{x\},\{x,y\})$, Case 3: $(\{y\},\{x\},\{y\})$, Case 4: $(\{y\},\{x,y\},\{x,y\})$, Case 5: $(\{y\},\{x,y\},\{y\})$, and Case 6: $(\{y\},\{y\},\{y\})$. For each case, we show that at least one of $f_P^k, f_B^k, f_A^k$ has class $\{y\}$. For cases 1, 3, and 6, this is obvious. For case 2, $\mathcal{L}^{k-1}[L^0,\ldots,L^{k-2}]$ is a singleton: $L_i^{k-1}$ : $f_3^{k-1} f_2^{k-1} f_1^{k-1}$ for all $i \in A$ and $L_i^{k-1}$ : $f_1^{k-1} f_3^{k-1} f_2^{k-1}$ for all $i \notin A$. Because $a < n/2$, we have $f_P^k(L^{k-1}) = \{f_1^{k-1}\}$, which means $C_{f_P^k} = \{y\}$. Case 4 is similarly shown. For case 5, we have $f_A^k(L^{k-1}) \subseteq \{f_1^{k-1}, f_3^{k-1}\}$ for all $L^{k-1} \in \mathcal{L}^{k-1}[L^0,\ldots,L^{k-1}]$.  ∎

**Sketch of the proof that** $\{f_{E_1}, f_{E_2}, f_{E_3}, f_B\}$ **satisfies** $p_{WC} \approx 1 - p_D$ **under EU and IAC.** Because of Lemma 1 and Lemma 2, the only nontrivial case is $f_1^1(L^0) = f_2^1(L^0) = \{x_1\}, f_3^1(L^0) = \{x_2\}$, and $f_4^1(L^0) = \{x_3\}$. If $f_1^k, f_2^k, f_3^k, f_4^k$ can drop $x_j \in X$ at some $k \in \mathbb{N}$, the proof is similar to $m = 3$ (using EU). Otherwise, for each $x = x_1, x_2, x_3$, at least one $f_x^k \in F^k$ exists such that $x \in f_x^k(L^{k-1})$ for all $L^{k-1} \in \mathcal{L}^{k-1}[L^0,\ldots,L^{k-2}] \ldots (\star)$. Since ties between $f_3^1$ and $f_4^1$ are negligible (when $n \to \infty$), each $f_x^k$ is assumed to be distinct. Furthermore, at least one individual is expected to exist who ranks $x_j$ last. Then, $f_A$ cannot be a $f_x$. Therefore, combinations of $\left(f_{x_1}^k, f_{x_2}^k, f_{x_3}^k\right)$ are the permutations of $f_{E_1}, f_{E_2}, f_B$. For each combination, we obtain a linear system of inequalities on the number of individuals who have specific preferences on $\{x_1, x_2, x_3\}$ (considering $k = 2, 3$). With Fourier-Motzkin elimination, we can find that at no case does $(\star)$ occur.  ∎

# References

1. Barbera, S., Bossert, W., Pattanaik, P.K.: Ranking sets of objects. In: Barbera, S., Hammond, P.J., Seidl, C. (eds.) Handbook of Utility Theory Volume II Extensions, pp. 893–977. Kluwer Academic Publishers, Dordrecht (2004)
2. Barbera, S., Jackson, M.: Choosing how to choose: Self-stable majority rules and constitutions. Q. J. Econ. **119**, 1011–1048 (2004)
3. Diss, M., Louichi, A., Merlin, V., Smaoui, H.: An example of probability computations under the IAC assumption: the stability of scoring rules. Math. Soc. Sci. **64**, 57–66 (2012)
4. Diss, M., Merlin, V.: On the stability of a triplet of scoring rules. Theor. Decis. **69**(2), 289–316 (2010)
5. Eğecioğlu, Ö.: Uniform generation of anonymous and neutral preference profiles for social choice rules. Monte Carlo Methods Appl. **15**(3), 241–255 (2009)
6. Endriss, U.: Sincerity and manipulation under approval voting. Theor. Decis. **74**(3), 335–355 (2013)
7. Gillet, R.: Collective indecision. Behav. Sci. **22**(6), 383–390 (1977)
8. Gillett, R.: The comparative likelihood of an equivocal outcome under the plurality, Condorcet, and Borda voting procedures. Public Choice **35**(4), 483–491 (1980)

9. Houy N.: A note on the impossibility of a set of constitutions stable at different levels. Mimeo (2004)
10. Koray, S.: Self-selective social choice functions verify arrow and Gibbard- Satterthwaite theorems. Econometrica **68**, 981–995 (2000)
11. Koray, S., Slinko, A.: Self-selective social choice functions. Soc. Choice Welf. **31**(1), 129–149 (2008)
12. Marchant, T.: The probability of ties with scoring methods: some results. Soc. Choice Welf. **18**(4), 709–735 (2001)
13. Merlin, V., Tataru, M., Valognes, F.: On the probability that all decision rules select the same winner. J. Math. Econ. **33**(2), 183–207 (2000)
14. Nurmi, H.: Discrepancies in the outcomes resulting from different voting schemes. Theor. Decis. **25**(2), 193–208 (1988)
15. Pritchard, G., Wilson, M.C.: Exact results on manipulability of positional voting rules. Soc. Choice Welf. **29**(3), 487–513 (2007)
16. Pritchard, G., Wilson, M.C.: Asymptotics of the minimum manipulating coalition size for positional voting rules under impartial culture behaviour. Math. Soc. Sci. **58**(1), 1–25 (2009)
17. Saari, D.G.: Millions of election outcomes from a single profile. Soc. Choice Welf. **9**(4), 277–306 (1992)
18. Saari, D.G., Tataru, M.M.: The likelihood of dubious election outcomes. Econ. Theor. **13**(2), 345–363 (1999)
19. Suzuki, T., Horita, M.: How to order the alternatives, rules and the rules to choose rules: when the endogenous procedural choice regresses. In: Kamiński, B., Kersten, G.E., Shakun, M.F., Szapiro, T. (eds.) GDN2015. LNBIP, vol. 218, pp. 47–59. Springer, Heidelberg (2015)

# Estimating Computational Models of Dynamic Decision Making from Transactional Data

James Brooks, David Mendonça, Xin Zhang$^{(\boxtimes)}$,
and Martha Grabowski

Rensselaer Polytechnic Institute, Troy, NY, USA
zhangx22@rpi.edu

**Abstract.** The goal of this work is to estimate and validate computational models of dynamic decision making against data on sequences of actual decisions made in naturalistic settings. While this paradigm has its roots in laboratory studies under controlled conditions, increasing instrumentation of operational environments is enabling parallel investigations in field settings. Here, decision processes associated with the dispatch of debris removal personnel and equipment are investigated using data from a series of tornado storms in the State of Alabama in 2011. A multi-faceted approach to model validation is presented, thereby illustrating how objective, operational data may be used to inform models of complete decision making processes.

**Keywords:** Dynamic decision making · Task environment · Decision models · Model validation

## 1 Introduction

Dynamic decision making is interdependent decision making that takes place in an environment that changes over time either due to previous actions of the decision maker or due to events that are uncontrollable by the decision maker [1]. It happens almost in all industries. For example, there are many dynamic systems in which resources are needed to be relocated to a network including mining, logging, and debris removal following natural disasters.

Estimating dynamic decision making models from these large complex systems not only gives a whole picture of the essential components which exist in the system (from work structure to human decision makers), but sheds light on how human cognition responses and interacts with the outside fast-changing environment. Knowing how decision makers would react dynamically in complex environments can lead to the development and improvement of the whole system. Simply modeling the human decision making behavior is not enough. A solid model on the system structure should be built first in order to further develop reasonable human decision making model. Recently, detailed transactional data from these systems is becoming available. However, these data are likely to be incomplete from the perspective of a system modeler. Thus, novel techniques are required in order to generate and validate reasonable dynamic decision making models.

© Springer International Publishing AG 2017
D. Bajwa et al. (Eds.): GDN 2016, LNBIP 274, pp. 57–68, 2017.
DOI: 10.1007/978-3-319-52624-9_5

Formulating dynamic decision making models based on transactional data requires a comprehensive understanding of the system in which the decision making process is taking place. In different dynamic decision making environments, the specific work structure models may be different, but the underlying human decision making behavior is similar. After attaining the estimated decision making model, the next essential step is to validate the model with real world data. If the model generates or predicts output similar to the actual data from the system, then the model is validated.

The goal of this work is to estimate computational models of human dynamic decision making from existing transactional data. These estimated models will provide insights on how human decision makers would change the decision they make when they continuously interact with the system. Previous work on building dynamic human decision making models and their validation is discussed in Sect. 2. A novel approach to estimating dynamic decision making models is introduced, and how these models can provide heuristics and insights on predicting decisions is illustrated in Sect. 3. Then the transactional data from debris removal operation in Alabama 2011 are used to validate the estimated models in Sect. 4. Finally in Sect. 5, the conclusion and future work are discussed.

## 2   Background

The structuring of dynamic decision making models is arguably one of the most important areas in studying human decision-making processes [2]. As decision making in dynamic systems requires an ability to deal with incomplete information, highly connected variables and changing environments over time, it has been argued that modeling of the system structure is essential to investigating human decision making behaviors in these dynamic systems [3].

Existing work on structuring the system model involves defining an intellectual process during which a problem situation is specifically constructed [4], conducting a search for the underlying structure or facts [5], and exploring various states of nature, options and attributes for evaluating options [6]. These methods of constructing system structure model provide their own definition and techniques for structuring dynamic decision making systems independently in a unique way [7]. For large complex dynamic decision making systems, an integration of some of these methods would suggest a better approach to constructing the underlying system structure.

Theories of modeling dynamic decision making include instance-based learning, recognition-primed decision making, and planned behavior. Instance-based learning theory tries to approach this problem by combining several human cognition characteristics as the decision making in dynamic environments involves learning [8]. Recognition-primed decision making model compares the decisions made by the decision makers and those imposed by the dynamic environments [9]. The theory of planned behavior models human decision making by studying perceived behavior control that relates to decision maker's beliefs and intentions [10]. These theories have laid a solid foundation for studying human decision making and encouraged new ideas of constructing decision making models.

Modeling frameworks for dynamic decision making include, in rough order of decreasing cognitive plausibility, various cognitive architectures [11], beliefs-desires-intentions [12], simple production systems [13], regression equations [14], and Kalman filters [15]. These models are typically generated through methods such as task analysis, interviews, questionnaires, or detailed behavior tracking. Other general methods of eliciting knowledge include critical decision method [16] and unstructured interview methods [17].

The proposed approach to modeling dynamic human decision making is to utilize the transactional record of an individual's decisions over time to estimate the best-fit parameters of quasi-rational models of decision making behavior [18]. Validation has been noted as a critical but often overlooked enterprise in modeling of dynamic systems. Several general validation methods have been proposed depending on the available system data [19]. Of interest for the present work is trace-driven validation in which historical input data is used to generate comparable simulated output data, and thus to support comparison of predicted vs. actual data.

# 3 Approach

When a decision maker is in a dynamic decision making environment, the decision maker will process the information obtained from the environment and then make the decision. Moreover, the decisions being made will eventually have an impact on the environment, which will then change the decision event and how decision maker makes future decisions. In order to estimate dynamic decision making models from field transactional data, a comprehensive understanding and modeling on the decision making environment is essential. The overall conceptual modeling process is shown in Fig. 1 as follow.

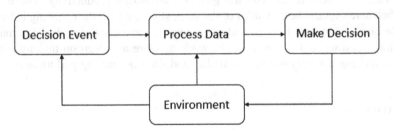

**Fig. 1.** Conceptual modeling process.

The approach proposed here consists of two steps: building the appropriate model for the decision making environment first, then estimating the dynamic decision making model embedded within this task environment. Due to the repeated nature and rapid pace of the scheduling that takes place in the domain of post-disaster debris removal, it is used as an application example to illustrate the proposed modeling approach.

## 3.1 Application Domain

Debris removal is a critical but understudied phase of disaster response and recovery [20]. Highly distributed team work is central to the execution of the debris removal mission [21]. As shown in Fig. 2, the debris field consists of a potentially wide variety of material distributed over the affected area. Debris is segregated and loaded onto trucks and trailers at designated pickup points by debris removal teams (DRTs). Upon leaving the loading site, trucks travel the road network to deliver loads to a temporary debris storage and reduction site (TDSR), where the size of the load is estimated visually as a percentage of the truck's total capacity by a human operator. A guarantee of payment (called a load ticket) is delivered to the truck driver, who then travels the road network to rejoin the DRT at the pickup point.

(a) Debris Field    (b) Debris Removal Team    (c) TDSR Site

**Fig. 2.** Activities in the debris removal mission.

The payment incentive structure is hypothesized to shape the decision process of subcontractors and their respective dispatchers. The number of hauling vehicles working on any team which delivers material from each remote site is under the control of the dispatcher. The task of the dispatcher is to dynamically allocate hauling vehicles (e.g., trucks) to these teams with the goal of maximizing productivity. The models described here capture the dynamics of the work structure (i.e., the queueing behavior) and the human dynamic decision making aspect (i.e., the dispatcher decision making behavior). As discussed below, both the work structure and decision making process must be addressed jointly in order to understand decision making performance.

## 3.2 Data

The data source is a database of load tickets of debris hauled during the mission to clean up after the tornadoes that devastated the state of Alabama in April of 2011. All load ticket data were obtained from the USACE (United States Army Corps of Engineers) in machine-readable format, then cleaned and verified via automated and manual methods [22]. Each record contains information regarding the pickup and drop-off locations, the amount of debris hauled, loading and unloading time stamps, a truck identifier, and other organizational identifiers. These organizational identifiers (namely, quality control identifier) allow one to infer each team change (reallocation) decision made within any day. Quality control personnel (QCs) are assigned to work teams on a one-to-one basis for any given day. Each data record contains a QC identifier and thus the team structure can be inferred from the data.

Once the teams are identified, relevant team characteristics must be determined. These are hypothesized to be mean haul distance, mean round-trip time, and team size. The mean round-trip time is determined by collecting the difference in loading times between adjacent trips for each truck within the team. Similarly, the mean haul distance is the mean for all trips delivered by a team, while team size is the number of trucks.

Thus, this transactional database provides detailed information on the dispatcher decision making process, including the time of the decision, the content of the decision, and most importantly the change of decisions (i.e. truck change among teams). It can serve as an access to observe human decision making behavior which continuously interacts with the environment. Moreover, it is of much value in terms of validating the estimated models.

### 3.3  Work Structure Model

In large complex resource allocating systems like debris removal operation, the systems may contain several interrelated queueing networks. In debris removal operations, the network has many parallel cycles in which several types of material are delivered from many remote sites to a fewer number of processing sites by some fixed number of hauling vehicles. Each processing site is assumed to handle some subset of the material types. The number of hauling vehicles working on any team which delivers material from each remote site is under the control of the dispatcher as shown in Fig. 3. The number of hauling vehicles being sent to each remote site is by $N_i(i = 1, 2, \ldots, n)$. Team and remote site will be used interchangeably throughout, i.e. the number of hauling vehicles being dispatched to each remote sites or processing center is dynamic and can be changed at any time by the dispatcher with the goal to maximize the profit.

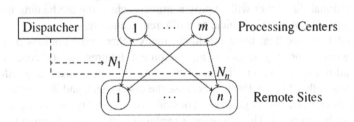

**Fig. 3.** Example parallel-cycle queueing network under control by a single dispatcher.

A discrete-event approach is taken for modeling queueing behavior of arriving customers at each of the nodes in the network (i.e., remote sites and processing centers), which are assumed to be single server queues without reneging, jockeying, or other non-standard customer behavior. The customers cycling in the queueing network represent hauling vehicles which are assumed to be non-autonomous in the sense that they obey the dispatcher's orders. While the remote site is determined for each customer by the dispatcher, the processing center is selected according to the material type for each load. It is assumed that the material type is randomly selected for each load according to some distribution. Each queue is further assumed to have an exponential service distribution. Finally, travel times between nodes in the network are assumed to be constant as well.

## 3.4   Dispatcher Decision Model

Based on the proposed work structure model, a model on the dispatcher who makes series of dynamic decisions throughout the whole operation can be attained. In debris removal operations, each dispatcher has been assigned a geographic region to clean. However, the number of trucks and their allocation between the ground crews are completely under the control of the dispatcher. Each company is paid according to the volume of debris delivered. Therefore, one can expect a rational economic model to be a reasonable model of the dynamic decision making process.

Due to the high complexity of the system, it is expected that complete knowledge of the system is not possible. As a result, let t and w be the vectors of estimated travel and wait times, respectively. Further, suppose that the per-volume payment can vary according to the pickup site. Let p be the vector of these per-volume payment amounts. Finally suppose the vehicle allocation can be described as a vector of allocated debris (vehicle) flows, a. Under the rational economic assumption, the dispatcher is expected to maximize the following function, where c is their average time cost of operation.

$$f(p, w, a) = \left[ \frac{p}{w + t} - c \right]^T a. \tag{1}$$

Based on the proposed economic model, when the wait time or travel time for a hauling vehicle is too long, or the payment is not high enough, in order to maximize the profit, the rational dispatcher will employ a strategy: changing the hauling distance of the vehicle, or allocating this vehicle to a different team, etc.

It should be noted that there are likely many other considerations a dispatcher considers when allocating trucks (e.g., equipment failures, safety concerns, other ground conditions). However, the available data do not contain any information regarding these other factors. Further, because the dispatchers, and information sources (i.e., team members), are cognitively constrained, variability in selecting the best decision is to be expected. This variation is captured in the model through the use of a Boltzmann distribution (Luce's choice axiom) in which the probability of selecting any one option is proportional to the value of the choice [23].

Further, it is assumed that decisions are made at random points in time, herein called decision points, and that each instance results in one and only one truck being transferred. Thus, if there are n teams, there are $n(n-1)$ possible decision options. The value of these options is quantified by the log ratio of the earning rates (the difference between payment rate and cost rate), denoted by $r_k$, for decision $k \in [1, n(n-1)]$. The probability of making decision $k$ is then given by

$$p(r_k) = \frac{e^{\frac{r_k}{T}}}{\sum_i e^{\frac{r_k}{T}}},\tag{2}$$

where the amount of stochastic variation allowed is controlled by the parameter T. A low parameter value gives highly rational (i.e. deterministic, optimal) behavior, while a high parameter value gives completely stochastic selection (i.e. all choices equally likely).

The dispatcher model execution then follows the flow shown in Fig. 4. The decision points are chosen randomly according to some inter-decision distribution. At each point, the estimates of cycle times and the known payments are then considered to make the decision.

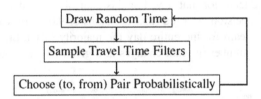

**Fig. 4.** Flow chart of dispatcher decision model execution.

These decisions are then communicated to the truck drivers, who alter their path accordingly after leaving the next processing center. To minimize any delay in executing the decision, eligible trucks are first sought at a processing center. If none can be found, the switching truck is chosen at random. In the case where a decision is selected in which the original team currently has no trucks assigned, then nothing is done and execution proceeds with the same allocation until the next decision point.

As a dispatcher's goal is to maximize the profit, the following heuristic can be generated from the proposed models: with less hauling distance, less round-trip time, and smaller team, the dispatcher gets more profit. Thus they are (either consciously or unconsciously) prone to making series of dynamic decisions to shorten the hauling distance and round-trip time, as well as decreasing the team size.

## 4 Validation

The models (work structure model and dispatcher decision model) are not able to be validated effectively as one large model. Thus, a multi-stage approach will be used, starting with portions of models which can be validated independently with field data.

### 4.1 Validation Method

The dispatcher decision model is validated against observed allocation decisions in the field data. A maximum likelihood approach is used in which the model parameters are chosen to maximize the total likelihood of all observed decisions. For the remaining validation procedures, the dispatcher decision model will be disabled, i.e. a fixed allocation will be used. A trace-based validation method is expected. Here, the system state can be fed into the dispatcher decision model and the resulting decisions can be compared to the actual decisions made in both quantitative and qualitative manners, e.g. by looking at a distribution of old and new characteristics.

## 4.2  Analysis

A portion of the data (the month of June) was used to validate the proposed models and its embedded heuristics. Team change decisions are evident in 13.43% of the truck-days available in the June data, while the remaining 86.57% of the time a truck stays with the same team for that day. For this analysis, only data which contains a single team change decision are used (833 decisions or 9.05% of team-days). A truck stays with the same team for the entire day the majority of the time (86.6%). A histogram showing the number of team changes in a given day is shown in Fig. 5.

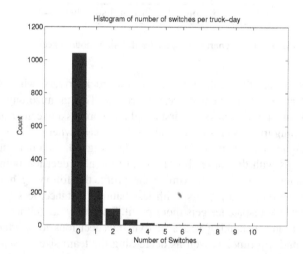

**Fig. 5.** Histogram of number of team switches (truck-day).

In Fig. 6, haul distance, round-trip travel time and team size are shown with the value from the original team on the x-axis and the value from the new team on the y-axis. Each decision is shown as one point in this scatter plot. These plots indicate the distribution of team characteristic changes in hauling distance, round-trip time, and team size. The unit-slope line shown represents no difference in the given measure between the original (first) and new (second) teams. The percentage of points falling below this line, implying improvement in the given measure, as shown in Table 1. All combinations of these indicators are shown in Table 2.

Based on Table 1, the number of decisions on team changes being analyzed is 1771. And approximately 81.31% of decisions are team characteristics changes related

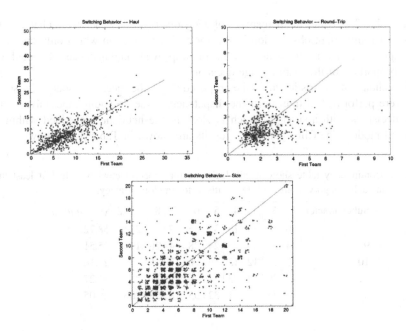

**Fig. 6.** Difference in haul distance (mi), round-trip time (h), and team size (number of trucks).

**Table 1.** Percentage of all team change decisions meeting each criteria for the new team.

| | |
|---|---|
| Number of decisions analyzed: | 1771 |
| Haul smaller: | 50.31% |
| Round-trip time smaller: | 42.80% |
| Team smaller: | 40.77% |
| Any of above: | 81.31% |

**Table 2.** Percentage of all team change decisions meeting each combination of criteria.

| Category | Description | Percentage of decisions |
|---|---|---|
| 1 | None | 18.69% |
| 2 | Haul only | 15.87% |
| 3 | Time only | 13.55% |
| 4 | Haul and time | 11.12% |
| 5 | Size only | 12.14% |
| 6 | Size and haul | 10.50% |
| 7 | Size and time | 5.31% |
| 8 | All | 12.82% |

to hauling distance, round-trip time and team sizes, with 50.31% changes on shortening hauling distances, 42.80% on shortening round-trip time and 40.77% on decreasing team sizes. These results also suggest that no one measure is indicative of the decision outcome.

As Table 2 provides the percentage of decisions regarding each team characteristic change, one important observation is that for both decisions in which only one factor was improved and those for which two factors improved, haul distance accounted for a larger proportion of the team change decisions.

To validate the heuristic within the proposed dispatcher decision model, Chi-square tests were performed to indicate if the dispatchers made decisions randomly, or they made decisions with a strategy which is close to the heuristic embedded within the proposed model. The Chi-square tests results are shown in Table 3 as follow.

**Table 3.** Contingency table showing all decisions for subcontractors who had at least three decisions in each category along with test statistic for random strategy.

| Subcontractor | 1 | 2 | 3 | 4 | 5 | 6 | 7 | 8 | Total | $x^2 - Statistic$ |
|---|---|---|---|---|---|---|---|---|---|---|
| 5 | 50 | 43 | 39 | 13 | 32 | 35 | 7 | 15 | 234 | 58.72 |
| 9 | 34 | 27 | 33 | 25 | 24 | 28 | 4 | 20 | 195 | 25.51 |
| 10 | 20 | 13 | 12 | 13 | 6 | 6 | 3 | 6 | 79 | 22.16 |
| 12 | 48 | 39 | 14 | 23 | 31 | 16 | 29 | 33 | 233 | 31.27 |
| 22 | 33 | 37 | 42 | 31 | 32 | 35 | 18 | 58 | 286 | 25.05 |

Based on the Chi-square tests, as all test statistics are larger than the critical value $x^2_{0.01,7} = 18.48$, then there is significant evidence that the subcontractors did not make those team characteristic change decisions randomly. They made the decisions with a strategy to maximize their profit, which includes shortening hauling distance, shortening round-trip time and decreasing team-sizes. Thus, the proposed models and the embedded heuristic are validated.

# 5  Conclusion

This work proposes a novel approach to estimating and validating computational dynamic decision making models using transactional data. Based upon the work structure model of a dynamic system, the estimated model can capture the embedded human decision making strategy. Thus, this approach contributes to the study of human cognition in dynamic decision making settings in the sense that it provides a new perspective of observing and modeling human decision making behavior.

In the field of debris removal mission, the two interrelated models (work structure model and dispatcher decision model) have been validated against actual data from a recent debris removal mission. The results indicate that the proposed models are able to capture the underlying decision making processes in complex resource allocating systems and to produce heuristics close to actual strategy used by dispatchers in debris removal operations. Thus, this modeling approach is useful, in terms of both formulating complex dynamic resource allocating problems and eliciting the human decision making processes. The computational and optimization models developed here would also be useful for planning purposes. This could include number and location of temporary debris reduction and storage sites, cost-benefit analysis of new equipment types, and other alternate operating configurations. Further, these models could be extended to include facility location decisions and used, along with debris field estimates, to predict expected mission duration, and benchmark system performance generally.

Future work includes an investigation of using the estimated and validated dynamic human decision making models to predict decisions, so that it can provide guidance for operations in the field. Another research direction is to view the transactional data from a different perspective, e.g. team decisions rather than individual decisions. In this case, the unit of analysis is team, and the modeling approach would be team-centric and process-oriented.

**Acknowledgments.** This material is based upon work supported by the National Science Foundation under Grant No. 1313589.

# References

1. Brehmer, B.: Dynamic decision making: human control of complex systems. Acta Psychol. (Amst) **81**(3), 211–241 (1992)
2. Mintzberg, H., Raisinghani, D., Theoret, A.: The structure of 'unstructured' decision processes. Adm. Sci. Q. **21**, 246–275 (1976)
3. Schmid, U., Ragni, M., Gonzalez, C., Funke, J.: The challenge of complexity for cognitive systems. Cogn. Syst. Res. **12**(3), 211–218 (2011)
4. Majone, G.: An anatomy of pitfalls. In: Pitfalls of Analysis, pp. 7–22 (1980)
5. Thierauf, R.: An introductory approach to operations research. J. Oper. Res. Soc. **30**(12), 1134 (1979)
6. Keller, L.R., Ho, J.L.: Decision problem structuring: generating options. IEEE Trans. Syst. Man Cybern. **18**(5), 715–728 (1988)
7. Corner, J., Buchanan, J., Henig, M.: Dynamic decision problem structuring. J. Multi-Criteria Decis. Anal. **10**(3), 129–141 (2001)
8. Gonzalez, C., Lerch, J.F., Lebiere, C.: Instance-based learning in dynamic decision making. Cogn. Sci. **27**(4), 591–635 (2003)
9. Klein, G.A.: Sources of Power: How People Make Decisions. MIT Press, Cambridge (1999)
10. Ajzen, I.: The theory of planned behavior. Organ. Behav. Hum. Decis. Process. **50**(2), 179–211 (1991)
11. Brooks, J.D., Wilson, N., Sun, R.: The effects of performance motivation: a computational exploration of a dynamic decision making task. In: Proceedings of the First International Conference on Brain-Mind, pp. 7–14 (2012)
12. Lee, S., Son, Y.-J., Jin, J.: An integrated human decision making model for evacuation scenarios under a BDI framework. ACM Trans. Model. Comput. Simul. **20**(4), 23 (2010)
13. Hattori, H., Nakajima, Y., Ishida, T.: Learning from humans: agent modeling with individual human behaviors. IEEE Trans. Syst. Man Cybern. Part A Syst. Hum. **41**(1), 1–9 (2011)
14. Crowder, R.M., Robinson, M., Hughes, H.P.N., Sim, Y.-W.: The development of an agent-based modeling framework for simulating engineering team work. IEEE Trans. Syst. Man Cybern. Part A Syst. Hum. **42**(6), 1425–1439 (2012)
15. Pattipati, K.R., Kleinman, D.L., Ephrath, A.R.: A dynamic decision model of human task selection performance. IEEE Trans. Syst. Man Cybern. **13**(3), 145–166 (1983)

16. Klein, G., Calderwood, R., Macgregor, D.: Critical decision method for eliciting knowledge. Syst. Man Cybern. IEEE Trans. **19**(3), 462–472 (1989)
17. Dutton, J.M.: Production scheduling—a behavioral model. Int. J. Prod. Res. **3**(1), 3–27 (1964)
18. Diehl, E., Sterman, J.D.: Effects of feedback complexity on dynamic decision making. Organ. Behav. Hum. Decis. Process. **62**(2), 198–215 (1995)
19. Kleijnen, J.P.C.: Verification and validation of simulation models. Eur. J. Oper. Res. **82**(1), 145–162 (1995)
20. Luther, L.: Disaster debris removal after hurricane Katrina: status and associated issues (2006)
21. Mendonça, D., Brooks, J.D., Grabowski, M.: Linking team composition to team performance: an application to postdisaster debris removal operations. IEEE Trans. Hum.-Mach. Syst. **44**(3), 315–325 (2014)
22. Brooks, J.D., Mendonça, D.: Simulating market effects on boundedly rational agents in control of the dynamic dispatching of actors in network-based operations. In: Proceedings of the 2013 Winter Simulation Conference: Simulation: Making Decisions in a Complex World, pp. 169–180 (2013)
23. Luce, R.D.: The choice axiom after twenty years. J. Math. Psychol. **15**(3), 215–233 (1977)

# Demystifying Facilitation: A New Approach to Investigating the Role of Facilitation in Group Decision Support Processes

Mike Yearworth[1]([⊠]) [iD] and Leroy White[2] [iD]

[1] Business School, University of Exeter, Exeter, UK
M.Yearworth@exeter.ac.uk
[2] Warwick Business School, University of Warwick, Coventry, UK
Leroy.White@wbs.ac.uk

**Abstract.** We believe there are still gaps in our knowledge of the facilitator role in group decision support processes and these must be "de-mystified" if use of these methods is to become more widespread. We use the design and analysis of online group model building to form a better understanding of the facilitator role. Our experimental configuration makes use of Group Explorer (Decision Explorer), configured to be delivered across the Internet in a distributed manner. The facilitator is thus no more or less visible in the workshop as any other participant. Data from a workshop is analysed and the findings discussed in relation to the following works; (i) Callon on translation, (ii) Hiltz et al. on the problem of animating methodology, and (iii) White and Taket's "death of the expert". We conclude by discussing one possible end-point of this work – the rise of a participant-led group decision support process model.

**Keywords:** Group decision support · Group model building · Journey making · Facilitation · Ethnomethodology · Problem Structuring Methods (PSMs) · Group Support Systems (GSS) · Group Decision and Negotiation (GDN)

## 1 Introduction

Debates about the processes of Group Decision and Negotiation (GDN) generally focus on methodology, expertise and facilitation, often independently, but sometimes conflated. But on the rare occasion where they are held just so far apart as to bring forth insights on the need to explore the ideas further, the comments and conclusions appear all too apparent. We therefore seem no further forward in our understanding of the practices and processes that should be adopted in pursuit of improved GDN performance than we were following Eden and Radford's seminal collection of studies on group decision support for strategic action [1]. Noting the failure of interventions in the realm of management practices, Eden and others encouraged academics and practitioners to be wary of dismissing such interventions on a matter of principle, portraying failure as one purely due to implementation that necessitated more contextualized and nuanced consideration of GDN practices from a hard setting to a soft [2–5].

© Springer International Publishing AG 2017
D. Bajwa et al. (Eds.): GDN 2016, LNBIP 274, pp. 69–86, 2017.
DOI: 10.1007/978-3-319-52624-9_6

We suggest, therefore, that there remains a research gap in terms of the need for theoretically informed empirical work to reflect the complexities of different processes for GDN; in other words, to employ more holistic approaches to process performance that reflect the many different demands that may be placed on a GDN intervention; and to review the complex relationships that may exist between GDN context and performance.

In particular, in this paper, we still see a large gap in our understanding and knowledge of the facilitator's role and that this must be understood and "de-mystified" as we consider the transition of GDN practices to a soft setting, as exemplified by the use of group model building in Problem Structuring Methods (PSMs). Our research focus is specifically on participant-to-facilitator interactions. Our context and opportunity arises from the need to facilitate stakeholder groups through a process of problem structuring where these groups are increasingly geographically dispersed. We base this on the evidence of requirements for four projects where the authors are either advising on the use of PSMs, or are directly involved as facilitators, where the staging of workshops with participants attending in person is proving difficult.

Building on the work of [6] we follow the idea of *"distributed interaction within a PSM process"*, but still see the workshop as an important component of the process, at least virtually. With the general improvement in the quality of network connections and collaboration tools, coupled with low-cost easy to access cloud-based compute infrastructure we believe the means for exploring this way of working is now technically feasible and methodologically justifiable, hence the reality of *distributed* Group Support Systems (GSS) as a means of implementing group model building in a PSM. Naturally, the distributed nature of stakeholder interaction e.g. in the case of a charity with stakeholders spread between the UK and Asia, is itself part of the problematic situation and we are sensitive of the fact that distributed interaction within the PSM process cannot be separated from this. The empirical work we report in this paper is an exploration of the issue of facilitation as we establish a working environment in which to conduct such online, virtual workshops. The data we analyse is collected from one of these online workshops where we have demonstrated the capability to problem owners in organisations we are working with and where the presenting issue is in fact the question of how to make this distributed engagement work.

We adopt an experimental setup that makes use of Group Explorer (Decision Explorer), a GSS that is based on the Journey Making methodology [7–9], but delivered across the Internet in a distributed manner. Consequently, the facilitator is no more or less visible and involved in the workshop as any other participant by way of its distributed nature. Our analytical technique is based on ethnomethodology [10], which has recently been used to good effect to understand the micro-process of decision-making in workshops (e.g. [11–14]). In so doing, we make the following contribution to the literature. First, we build on the foundations established by [6], to form a better understanding of the role of the facilitator in this type of setting. In particular, our attention is focussed specifically on participant-to-facilitator interactions. We theoretically position our work in relation to the following: (i) the work of Callon [15] on translation and specifically how facilitation addresses the questions of problematisation and interessement [16], (ii) the work of Hiltz et al. [17] on distributed GSS and the problem of animating methodology, and (iii) the "death of the expert" [18].

This perspective enables us to take a broader and nuanced view of expertise, which gets to the heart of investigating the role of the facilitator as an expert in methodology. Finally, data from one of the workshops organised to demonstrate technical and methodological feasibility in this distributed manner is analysed and the findings discussed in relation to our theoretical expectations. In particular we examine the question of the possible demise of the expert facilitator and the rise of a technology enabled and participant-led group decision support process model.

The remainder of our paper unfolds as follows. First, we review the literature on facilitation in GDN, delineating the dimensions of facilitation, explain our theoretical underpinning, and then bring the two ideas together in developing our theoretical model. Second, we present the data and method we employ. Third, we present and discuss our results. In our final section we highlight our contribution to extant literature and suggest implications.

## 2 Literature Review on Facilitation

Classic work on facilitation follows the seminal work by [19]. Here the concern was on the facilitator as the 'helpful intervener'. Here, the intervener strives to improve group dynamics and decision making or provide a learning environment to help participants gain confidence of an interpersonal nature in order to help them transform the patterns of communication. Indeed, much of the work on GDN focuses on the facilitator that strives to influence situations toward desired goals in the human activity systems in which they intervene. Here the facilitator attempts to balance what is to be done with how it will be achieved; see for example [4, 6, 20–26].

The role we are defining for the facilitator in this work is somewhat different. We begin to problematize the role as follows using group model building in a PSM engagement as the focus of our group decision support process. The question of facilitation seems to be situated within existing PSM practice, so the facilitator as a role is already there, has always been there in the process, and always originating from a *methodological* root. We imagine the written accounts of methodology as a product of first-hand experience in facilitating the methodology. There are many published methodologists, but they are all also practitioners. Is it therefore possible to theorise about PSMs without extensive practice-based experience? Problematisation seems to require us to break the bond between the roles of methodologist and facilitator and place our focus on deconstructing the latter; the thoughts of the methodologist are largely what we know already from the literature.

We can problematize the role in three ways:

1. Through isolating the facilitator by making them as on-par with a participant as possible, creating a laboratory to study facilitator interactions (online method);
2. By analysing case studies divorced from codified PSM methodology (and therefore the associated methodologist) i.e. deconstructing a non-codified case to tease-out the facilitation role (if any) (non-codified case method);
3. By distilling the essence of the facilitator role from the bulk of the PSM and GDN literature. To a certain limited extent this has been done in our literature review

here, but there is perhaps some further concentration that could be performed to tease-out nuances. However, it is unlikely to produce anything exceptional by way of results (literature method).

To some, the role of the facilitator seems tangled with questions of leadership and, worse, the notion of systems thinking [27, 28]. Not as anything precisely defined, and associated with any particular methodology, but as a quality of an individual that uniquely sets them apart to take on the role of facilitator when dealing with the sort of messy problems that PSM engagements are designed. We suggest that this is an unprofitable line of enquiry as it is unlikely to lead to any widely useful contribution. Our focus therefore remains with the performative i.e. what are facilitators actually doing when they facilitate a PSM engagement, and therefore preserve our theoretical underpinnings in ANT/Mangle [12]. Whilst it would be interesting to explore the analysis of the facilitator role in non-codified PSM use, precisely because it would be an analysis of facilitation by a non-methodologist in a PSM-like setting, we defer this to future work.

## 3   Theory

Recent focus drawing on pioneering work of Keys on the sociology of scientific knowledge [29–32], and recently work by OR scholars drawing on socio-materiality, particularly the works of Latour [16, 33, 34] and Pickering's Mangle [34, 35] attention has been paid to important agendas regarding theory, behaviour and outcomes pertaining to (particularly Soft) OR processes, including facilitation. We note that these studies have recognized that interventions are both temporally enacted affairs and concerned with becoming coordinated practices through the performance of using models as objects, but the studies are not adequate in addressing in full issues relating to facilitation in interventions. Therefore, some significant methodological and epistemological challenges remain [16, 36–38]. Relevant to our work on facilitation is an extenuation of socio-materiality from Callon [39]. He outlines a number of themes which we feel are relevant to our study, in particular, the Co-Production of Knowledge Model (CKM). In the CKM he recognises a persistently enriched contest between the production of standardised knowledge and knowledge that takes into account the complexity of local circumstances or context [39]. In the space between the two is the problem of facilitation. Callon's CKM notes that the typical mapping of the problem structuring onto an expert–lay divide – in which experts possess expertise and participants possess local knowledge that can challenge assumptions made by those applying expertise to particular contexts. The experts as facilitators do not capture the capacity of participants to be involved in all elements of knowledge production. Nor does it challenge understandings of a problem that may be as highly differentiated as those held by the expert. In the same way that experts question their understandings through practice, so must the participants. Thus, under CKM, knowledge is co-produced through a process of dynamic, collective learning involving those for whom an issue is of particular concern, whether as a result of their expertise, their personal position with respect to the problem at hand, or their personal experience of the problem. This explicitly recognises more socially distributed, autonomous and diverse forms of collective problem structuring [6, 40]. Problem structuring is no longer a property of expertise [18], and the knowledge it produces is no longer accorded special privilege over other knowledge. This process does not eliminate the need for the involvement of expertise, rather it removes its privilege and emphasises that it is, on its own, insufficient.

Callon's CKM is useful in capturing a theme central to debates over expertise in decision-making. Expertise is more widely distributed than many might imagine. The question becomes how to mobilise and to diversify that expertise and what happens to the expertise of the facilitator during this mobilisation. By addressing this question new kinds of understanding may be generated that unsettle the taken-for-granted aspect of problem solving. Here, we introduce an experiment that explores distribution of expertise/facilitation to other people, things and places. To understand why this distribution of expertise is different.

## 4   Constructing the Experimental Setup

A standard Group Explorer installation was repurposed for deployment in a cloud-computing environment. Group Explorer provides a 'wrapper' to the Decision Explorer software – which is *"designed to record, analyze and present qualitative data argumentation relating to strategic policy issues and modelled as cognitive maps"* [8] – such that multiple participants can share in the development of a cognitive map in a facilitated workshop. The wrapper provides a web-based interface to participants and also manages the various phases of model development and the storage of data about its dynamic development to support the sort of analysis presented later in this paper. A sketch of the installation and deployment process for the Microsoft Azure cloud computing service is described in Appendix A. The motivation for choosing this type of internet-based hardware platform was to address the following needs:

1. To facilitate remote connection to the Group Explorer environment from any geographic location without having to deal with organisational firewalls and access limitations;
2. To avoid procuring and maintaining hardware, thus shifting costs from capital to operational, consistent with many organisations' strategies to migrate their IT infrastructure to cloud services;
3. To instantiate the software environment only when required for a workshop, thus further pushing operational expenses as low as possible by making best use of the pay-for-use model of the service provider;
4. To enable migration of configured Virtual Machines (VMs) to higher processor and memory specification hardware should there be a need for increasing performance. The management and configuration tools from the service provider are specifically designed to monitor for performance issues and enable migration.

A single online collaboration tool was used to provide both the audio conferencing capability and the means to share the screen of the computer hosting the Decision Explorer modelling software. Feasibility was tested with both TeamViewer and Citrix

GoTo Meeting products. For the feasibility workshops reported in this paper the usual two facilitation roles of a Group Explorer workshop were replaced by a single facilitator, who was both facilitating the group and also controlling the modelling using Decision Explorer.

## 4.1   Conducting Experiments

The experiments reported in this paper were conducted with the main purpose of (i) establishing the technical feasibility of conducting Group Explorer workshops in a distributed online setting, and (ii) furnishing sufficient data to begin to analyse the facilitator role. Having agreed to take part in the testing of an online workshop the *technical* means to join the workshop was managed carefully with the clients. To help in the process of demonstrating feasibility to clients three documents were prepared and circulated to participants a few days before the workshop was due to take place. In addition to a data collection permission form, based on a standard template we use for normal workshops, we also provided a document with detailed instructions about how to join the meeting online, including what to do if technical problems are encountered, and a document describing an online connection *etiquette*, designed to deal with mitigating problems with dropped or failed connections. Note that due to the nature of the data recording process for analysis purposes any participant who feels unable to give consent is excluded from participating in the workshop. The online connection etiquette guide is shown in Appendix B, redacted to remove client identity and phone numbers.

The experimental setup is complicated with many procedural steps required to make sure all of the components are working correctly prior to the workshop start time. Consequently, a checklist was developed for use by the facilitator to orchestrate the workshop setup and an actual example is shown in Appendix C. Refining this checklist over time as experience has developed has also led to a realisation of the steps in the instantiation of an online PSM environment that could be automated in the future. A question we return to later.

## 4.2   Data Collection

Our approach is based on recording and analysing participant-facilitator interaction during an online PSM workshop. The data consist of audio transcripts and the Excel Spreadsheet created from the SQL Server Export Wizard report generated from the workshop data held in the Group Explorer database. The two datasets are linked together by timestamps.

# 5   Data Analysis

Three online group model building workshops to demonstrate capability were held as follows:

1. 2–5<sup>th</sup> October 2015. Initial experimentation between authors addressing the issue *"Making Group Explorer usable in a distributed mode"*.
2. 23<sup>rd</sup> October 2015. Bristol-based charity with a stakeholder group spread between the UK and Asia. Feasibility workshop addressed the issue of *"Defining the effectiveness for a charitable sustainable energy programme"*.
3. 27<sup>th</sup> October 2015. InnovateUK/NERC funded project to explore the impact of adverse climate events on the delivery of health services across a UK city. Feasibility workshop addressed the issue of *"Improving the resilience of healthcare provision in Bristol to extreme weather events"*.

The first workshop was focussed on the issue of <*Making Group Explorer usable in a distributed mode*> and where it first became apparent that the experimental setup was providing a means of precisely examining what it was that the facilitator was doing during a group model building workshop[1]. The fact that the facilitator was now connected to Group Explorer in much the same way as a workshop participant meant that facilitator-participant interaction was as open to examination from the data as any other. Whilst the research focus on facilitation emerged from this first workshop and thus set the focus for data collection in the second and third workshops, at the same time these were still addressing the ostensible purpose of testing the technical feasibility of the online setting with clients.

The model from the first workshop was used to distinguish those aspects of <*Making Group Explorer usable in a distributed mode*> that were technical and/or methodological in nature from those relating solely to facilitation. The audio data from the third workshop was used to demonstrate the transition from facilitator led participation to a phase where the participants were able to model without facilitator intervention. The data from the second workshop are not analysed here, but it does represent the first distributed use of Group Explorer with a client and the lessons learned informed the setup for the third workshop.

The second and third workshops were deemed successful in the sense of warranting the conclusion that the online capability was *operational*, that having demonstrated feasibility the approach could now be used for future client workshops. We have deferred the question of evaluating the online approach, using techniques such as described by [38, 41], to further work.

## 5.1 Distinguishing Technology and Methodology from Facilitation

The model that emerged from the first workshop is shown in Fig. 1.

---

[1] The initial issue explored was how to support asynchronous modelling too. The Group Explorer setup was left running all weekend, hence the date range, to enable issues to be added and connected in the model without the facilitator being present at the client console. Whilst it worked technically, it was decided fairly quickly that this mode of working would not be explored further. However, we already have requirements for distributed workshops that will involve stakeholders separated by many time zones and the methodological issues raised by periods of un-facilitated participation could thus be investigated with the same experimental setup.

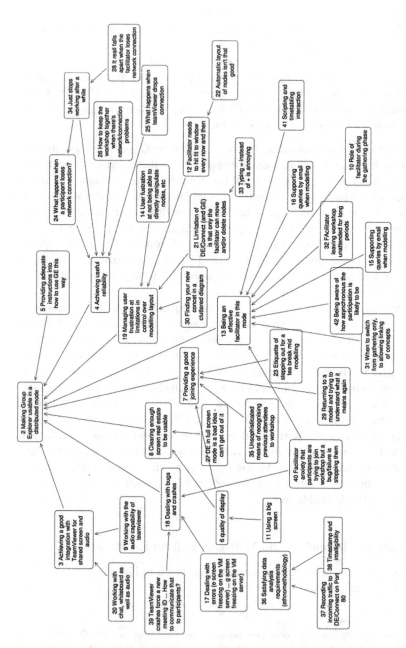

**Fig. 1.** Making group explorer usable in an online mode.

Themes emerging from this initial workshop can be broken down into the following categories:

1. Instructing participants in the use of the online systems (Group Explorer and the screen sharing and audio conferencing software);
2. Getting participants used to the way in which the modelling is designed to work e.g. entering and linking concepts;
3. Dealing with issues of poor and broken connections and technical limitations e.g. not seeing concepts immediately appear on the shared screen due to delays in providing a good layout for the participants;
4. Managing the process of participants modelling, enacting the script.

Themes 1–3 are mainly procedural arising from the technology, and methodology indirectly, and are thus candidates for automation and/or provision of better documentation to participants in the future. Theme 4 is the essential process of facilitation that we are trying to examine.

## 5.2   Detecting Transitions

The audio recording from the third workshop was analysed to demonstrate the feasibility of the approach for more detailed future analyses of facilitator-participant interaction and is shown in Table 1. The focus of the data presented here was the phase leading up to the first transition, from the workshop being facilitator led to one where the participants were able to focus on modelling from their own position of expertise without the facilitator's intervention. In the interests of space, the data and analysis of subsequent transitions is not presented.

**Table 1.** Data from the initial phase of the third workshop up to the first transition.

| $N_{event}$ | $T_{start}$ | Description |
|---|---|---|
| 1 | 0:00 | Facilitator is greeting participants as they appear on the audio conference and dealing with questions. One participant asks if there is enough time to *"go and make a cup of tea"* before the start, which they then proceed to do |
| 2 | 9:16 | Facilitator introducing the purpose of the workshop. Explaining something about PSMs generally, group model building and an overview of the technical means of how the workshop is being delivered. This is described in contrast to how the workshop would have been delivered in a conventional setting with the participants attending in person. During this time the preliminary model is being displayed via the shared-screen facility. After the preliminaries the facilitator explains the semantics of the model. In this case the 'blobs' are being interpreted as processes in a Hierarchical Process Model (HPM) and the 'links' as meaning 'part-of' relationships [42]. The model can thus be read as a system to achieve the purpose of *"improving the resilience of healthcare provision in Bristol to extreme weather events"*, much like a Purposeful Activity Systems (PAS) model in Soft Systems Methodology (SSM) [43]. The facilitator explains a simple linguistic game to constrain the verbs to gerunds, an important feature of HPM. During this time there are no interjections from the participants |

*(continued)*

**Table 1.** (*continued*)

| N<sub>event</sub> | T<sub>start</sub> | Description |
|---|---|---|
| 3 | 14:19 | A participant comments that their connection dropped for about two minutes "*I may have clicked the wrong thing*". The facilitator has not noticed the absence or anything amiss with the connection to the conferencing software and reassures the participant that they are 'back' in the meeting |
| 4 | 15:14 | Facilitator returns to the explanation described above |
| 5 | 16:01 | Facilitator now returns to explaining the preliminary model |
| 6 | 17:29 | Facilitator opens a new View in Decision Explorer to show a new process being created and explains how to use the web interface to Group Explorer to add new processes to the model. The facilitator starts by adding a new process via the Decision Explorer console on Public "*you should see a new process there, 17?*". Confirmed by a participant. "*Give me your ideas about how we can do this*", and reminds participants to play the gerund language game |
| 7 | 18:48 | Participant: "*Have you got mine there?*". Facilitator looks "*I can't see it at the moment*" and resizes the display to bring the new process into view |
| 8 | 18:55 | Participant "*Can you put up a view of the bigger diagram please*" – wants to see the original model. Facilitator switches display. Checking that the new process has been 'received' |
| 9 | 19:44 | Facilitator then switches View back |
| 10 | 20:04 | Participants first start reacting to each others' inputs to the model |
| 11 | 20:45 | Facilitator says "*OK, yes*" then is followed by a period of silence (keyboard noise heard) as the participants add processes to the model |
| 12 | 21:35 | Facilitator breaks silence by saying "*OK, this is all good stuff*" |

The start time corresponds to the announced meeting *connect* time, 15 min before the workshop was due to start. As can be seen below, 9 m:16 s of the 15 min allowed were required to establish attendance and connectivity.

## 6 Conclusions

Our work has focussed attention on shedding more light onto a subject that has remained equivocal. The process of developing an online group model building capability for projects with widely distributed stakeholders has given us an experimental framework to investigate the problem of facilitation at a micro-level. The attention to practical development of capability that could entail the decentring of the facilitator avoids the trap or descent into the purely critical and keeps the work at an empirical level.

The viewpoint piece [44] suggests that the development of group decision support has been by a number of '*gurus*' and reflects on their legacy and succession. As pointed out in the literature review this status of guru is associated with the *combined* role of facilitator and methodologist, although it is mainly knowledge about the latter that is reported; the healthiness of the field is evidenced by the continual development of methodology without much or any reference to the role of facilitation. Recent work by [45] provides hard data that can be used to refute any suggestion of stagnation "*When combined with other recent survey evidence, the use of PSMs and Business Analytics is apparently extending the scope of OR practice*", but the question remains whether these hints of a problem emerged because of something lacking in the area of facilitation, or more specifically in the facilitator as the *sine qua non* of methodological knowledge.

Our review of theory suggests that the role of facilitator, as *purveyor* of methodology in decision making engagements, is just another form of expertise that can be critiqued and potentially decentred from the essential business of group decision support. Our preliminary experiments have been light on methodology, both in terms of explanation to clients or in anything particularly creative in methodological design. The use of Group Explorer with a simple modification to the conventional use of Decision Explorer, coupled with its delivery online via a cloud service and with the workshop glued together by a reasonably sophisticated audio conferencing and screen sharing system provided a lot of the *scaffolding* for the group model building. In effect by implementing a distributed GSS that could be considered pre-packaged and largely separate from the process of facilitation. However, from the point of view of Callon on translation [15], and specifically how facilitation addresses the questions of problematisation and interessement [16], it was still the facilitator that initiated the workshops and who was essential on the conference call to explain how the process would work.

In the extract presented in Table 1 it is not until 20 min into the workshop that the facilitator stops being the expert in methodology and steps back to allow the participants to get on with the process of engaging with the problem. With regard to Callon's CKM we see that at this point the facilitator has been able to momentarily relinquish the expert role and allow the participants to be the experts in what they know. A translation where one sort of situated expertise (facilitator/facilitation) is transferred to another (participant/problem domain). The time spent up to this point was taken up by the facilitator translating expertise in methodology into practical explanation of process so that the participants could use it to enable their own expertise to become visible.

With regard to the work of Hiltz et al. on distributed GSS and the problem of animating methodology [17] our distributed group modelling capability is clearly not autonomous. The scaffolding might be there to enable self-animation on the part of the participants, but there was nothing in the preliminary guidelines that were circulated prior to the workshop that suggested participants could begin to model without the facilitator giving permission. Perhaps if the same group were to convene online in a subsequent workshop they might. However, even if Group Explorer had been started up in Gathering mode, the Chauffeur component of Group Explorer still requires a facilitator with access to the Chauffeur console to manually change configuration from Gathering to Preferencing to Voting. We can ask the question of whether a briefing note on the modelling process and some visual clues provided by a modified Group Explorer software itself would have been enough to get the stakeholder group modelling without the facilitator; but the question of who would have instigated the online workshop still remains. The question of animation, and particularly initiation and transition, is crucial to unpicking how a methodology plays-out in a group workshop and further analysis is required to fully understand this. Whilst we appreciate this would help us to improve group decision support processes generally, and is a worthwhile and perhaps necessary task, we also admit to the following agenda inspired by ideas of the *"death of the expert"* [18]. What if through further research we could understand the role of facilitator sufficiently well so that it could be *coded* into a software platform like Group Explorer? Rather than being puzzled by the question of whether a PSM engagement functioned because of the skill of the facilitator or because of a property of the methodology we would have sufficiently separated the two to gain clarity that the question of function could be investigated solely as a property of methodology. Although of course begging the question as to meaning of *function*. For the purposes of current experiments and future work our meaning is simply that of whether the group decision support process started at all and led to decisions being made.

We acknowledge the limitations in our work. Our analysis centres on the methodological, procedural and expert role of the facilitator, especially as initiator of process and enabler of lay expertise, mainly from the broad perspective of ANT. This has been at the expense of detailed micro-analysis using theories of behaviour such as Activity Theory [11]; however this is further work that can be carried out now that the experimental framework has been made operational and the method of data collection simplified. An additional strand of work we envisage is to return to the question of facilitator as initiator and how this role comes about, and an examination of the trust that must come into being in the relationship between the client and the facilitator.

To conclude, in our new experimental setting the facilitator has been literally decentred, the visual clues of being the centre of attention in the workshop have been removed and the facilitator is just another voice on the conference call. What if the audio cues could be replaced by software cues, perhaps supported by rule engine? This is speculation and perhaps where the development strands of GSS and PSM come together in a general group decision support process, but further understanding and de-mystification of the facilitator role may open the door to a proliferation of PSM/GSS application platforms. Whilst this may be technically feasible, this speculation brings us back to the essential puzzle of a PSM engagement; the initial problematisation and interessement [16]. Is this at all possible without a facilitator regardless of the properties of methodology and technical enablement of stakeholders?

**Acknowledgments.** We would like to thank the four anonymous reviewers for their useful review of our paper. We are indebted to Colin Eden for providing Group Explorer software and his patience, and that of Iain Small, in supporting a very non standard installation. We would also like to thank Paul Shabajee who helped with the first online workshop and provided input based on experience in investigating the user requirements for the STEEP collaborative stakeholder engagement platform (https://tools.smartsteep.eu/wiki/STEEP_Platform). We would also like to thank Alberto Franco and Ashley Carreras for their tutorial on Group Explorer at Euro 2015. This work was supported in part by (i) the EPSRC funded Industrial Doctorate Centre in Systems (Grant EP/G037353/1), (ii) the Innovate UK/NERC project Healthy Resilient Cities: Building a Business Case for Adaptation (Grant NE/N007360/1) and (iii) the EU FP7-ENERGY-SMARTCITIES-2012 (314277) project STEEP (Systems Thinking for Comprehensive City Efficient Energy Planning).

## Appendix A Installation Notes for Group Explorer in MS Azure

These notes are designed to help installing Group Explorer in Microsoft's Azure cloud computing environment using Windows Server 2008 R2 SP1 Virtual Machines (VMs) and refers to Group Explorer V2.1 User's Guide v2.1.3 and install files PublicSetup-v2.exe dated 28th March 2013 and ChauffeurSetup-v2.exe dated 9th December 2012.

**Part 1 – Creating the VMs, network, and assigning correct IP addresses**

- Create a suitable Microsoft Azure subscription
- Create a virtual network
  - Name: netnameXXXX
  - start address 192.168.0.0
  - CIDR/24(251) - creates a submask of 255.255.255.0
- Create the first virtual machine to host the Public computer
  - Compute->Virtual Machine->From Gallery->Windows Server 2008 R2 SP1
  - Name: PublicXXXX (whatever is needed to guarantee a unique name)
  - Region/Affinity: netnameXXXX
  - Endpoint: HTTP
  - Endpoint: GroupExplorerXXXX 8085
- Download the RDP connection file for PublicXXXX
- Connect to PublicXXXX
- Install dropbox
- Install Azure Powershell
- Shutdown Public
- Create the second virtual machine to host the Chauffeur computer
  - Compute->Virtual Machine->From Gallery->Windows Server 2008 R2 SP1
  - Name: ChauffeurXXXX (whatever is needed to guarantee a unique name)
  - Region/Affinity: netnameXXXX
  - Endpoint: HTTP
  - Endpoint: GroupExplorerXXXX 8085
- Download the RDP connection file
- Install dropbox
- Install Azure Powershell
- get credentials
- Configure IP address of PublicXXXX using Azure Powershell
- Shutdown ChauffeurXXXX
- Startup Public
- Check IP address of PublicXXXX
- Connect PublicXXXX
- Configure IP address of ChauffeurXXXX using Azure Powershell
- Startup ChauffeurXXXX
- Check IP address of ChauffeurXXXX

**Part 2 – Installing Public**

- On PublicXXXX
- Find the SQL Server 2008 R2 installer download page on the Microsoft website
- Download the installation file SQLEXPR_x86_ENU.EXE
- Start the SQL Server 2008 R2 install process by running SQLEXPR_x86_ENU.EXE
- Kill the install process to preserve extracted distribution
- Find and copy the extracted distribution tree for SQL Server 2008 R2 to Downloads
- Start Public Install as per the Group Explorer Install Manual

- When Group Explorer Installer starts the SQL Server installer change location of source to Downloaded file
- Let Public install finish

### Part 3 – Installing Chauffeur

- Login to ChauffeurXXXX
- Start Chauffeur Install as per the Group Explorer Install Manual
- Let Chauffeur install finish

# Appendix B Joining Instructions for an Online Meeting

 University of BRISTOL

### Online Workshop Etiquette

Dr Mike Yearworth
16th November 2015

#### Joining the meeting

1) On the day of the workshop you will be sent an email with the subject "ONLINE WORKSHOP: organisation | time". The email will contain two links. The message will look like:

> 15 minutes before the workshop is due to start please join the conferencing system by clicking on this link: http://go.teamviewer.com/v10/mXXXXXXXX
> Meeting ID: mXX-XXX-XXX

> If you have any problems connecting please call Mike on +44XXXXXXXXX

> At the time of the workshop please click on this link:
> http://chauffeur.cloudapp.net/groupexplorerconsole/

2) If anything happens that is making it difficult or impossible to continue participating in the workshop

> Please announce over the audio channel. This is experimental work so please let all the participants know what the problem is

3) Please use the private chat facility in TeamViewer sparingly, if at all.

> Ideally all communication should be mediated via the facilitator and/or model. Note that the chat channel is also recorded as part of the workshop.

#### Technical problems

1) If your internet connection drops

> Send an SMS to Mike on +44XXXXXXXXX saying who you are, that your connection has dropped, and an estimate of how long it will take to re-establish a connection and re-join the workshop

2) If TeamViewer doesn't connect to the meeting

> Send an SMS to Mike with a summary of the problem

3) If TeamViewer drops the meeting connection

> Attempt to re-join using the teamViewer meeting ID provided on the day of the meeting. If this doesn't work send an SMS to Mike with a summary of the problem

4) If Group Explorer is not allowing you to connect

> Let everyone know over the audio channel of TeamViewer

#### Data collection

1) See the separate document "Permission for workshop data collection."

> The audio channel and any chats between participants will be recorded for analysis and publication purposes. Note that no individual will be identified in any published work. A permission form will have been sent before attending the workshop. Anyone who has not agreed to these recording requirements will not have been sent a Team Viewer meeting ID number.

# Appendix C Online Meeting Configuration Checklist

**University of BRISTOL**

**Online Workshop Checklist**

Workshop Client: _CSE_
Date/Time: _29th January 2PM_
Meeting ID: _312-525-197_

| Item | Notes | Time/Value |
|---|---|---|
| **Approximately 6 hours before workshop...** | | |
| Startup GroupExplorer VMs | $ ./Public.sh then $./Chauffeur.sh | 07:57 |
| Check status | $ ./Status.sh | 08:02 |
| Check current balance | https://account.windowsazure.com/Subscriptions | £2:45 |
| Connect to Public using RDC | | 8:49 |
| Connect to Chauffeur using RDC | | 8:49 |
| Check for updates & install/restart | Chauffeur: none    Public: none | 8:50 |
| Create workshop dropbox | CSE-20160129 | 8:51 |
| Start GoToMeeting on Laptop | 312-525-197 | 8:52 |
| Copy GoToMeeting ID number | | 9:01 |
| Start GoToMeeting on Public | Use Public:Decision Explorer as participant name | 9:04 |
| Start Decision Explorer on Public | Load correct model for workshop, early joiners will see this model | 9:04 |
| Transfer meeting host from Laptop to Public | Check that workshop model is shared OK | 9:05 |
| Disconnect RDC clients from VMs | | 9:10 |
| Disconnect Laptop from meeting | | |
| Email meeting participants | | 9:10 |
| **Approximately 30 minutes before workshop...** | | |
| Check status | $ ./Status.sh | 13:24 |
| Re-connect Laptop to meeting | | |
| Connect to Public using RDC | | 13:24 |
| Connect to Chauffeur using RDC | | 13:24 |
| Kill Decision Explorer | | 13:32 |
| Start GroupExplorer components* on Public and Chauffeur | 1)Public:GE Remote Service, 2)Public:Decision Explorer, 3)Chauffeur:Group Explorer, 4)Public: GE Public | 13:32 |
| Load correct model for workshop | | 13:32 |
| **Immediately before workshop start...** | | |
| Start GoToMeeting recorder | Record screen and audio | 13:44 |
| **After workshop cleanup...** | | |
| Stop GoToMeeting recording | | 15:03 |
| Convert GoToMeeting AV file and move to workshop dropbox | CSE-20160129.MP4 | 15:07 |
| Stop Group Explorer components | | 15:03 |
| Export Chauffeur log file to workshop dropbox | Chauffeur-CSE-20160129.txt | 15:08 |
| Export SQLServer database on Public and save report to workshop dropbox | CSE-20190129.xls | 15:16 |
| Save Decision Explorer model file to workshop dropbox | CSE-Model1.MDL | 15:10 |
| Disconnect RDCs from VMs | waiting for MP4 upload to dropbox... | 15:35 |
| Shutdown Public VM | $ ./ShutdownPublic.sh | 16:09 |
| Shutdown Chauffeur VM | $ ./ShutdownChauffeur.sh | 15:56 |
| Check status | $ ./Status.sh | 16:12 |
| Check current balance | https://account.windowsazure.com/Subscriptions | £7:48 |
| **About 3 hours after workshop...** | | |
| Check VM usage stats | https://manage.windowsazure.com | 7:43 |
| Check current balance | https://account.windowsazure.com/Subscriptions | 7:45 |

* Group Explorer v2.1 User's Guide v2.1.3
Dr Mike Yearworth
25th January 2016

# References

1. Eden, C.E., Radford, J.E.: Tackling Strategic Problems: The Role of Group Decision Support. SAGE, London (1990)
2. Rosenhead, J.: What's the problem? An introduction to problem structuring methods. Interfaces **26**, 117–131 (1996)
3. Eden, C., Ackermann, F.: Where next for problem structuring methods. J. Oper. Res. Soc. **57**, 766–768 (2006)
4. Bell, S., Morse, S.: Groups and facilitators within problem structuring processes. J. Oper. Res. Soc. **64**, 959–972 (2013)
5. Joldersma, C., Roelofs, E.: The impact of soft OR-methods on problem structuring. Eur. J. Oper. Res. **152**, 696–708 (2004)
6. Morton, A., Ackermann, F., Belton, V.: Problem structuring without workshops? Experiences with distributed interaction within a PSM process. J. Oper. Res. Soc. **58**, 547–556 (2007)
7. Montibeller, G., Shaw, D., Westcombe, M.: Using decision support systems to facilitate the social process of knowledge management. Knowl. Manag. Res. Pract. **4**, 125–137 (2006)
8. Eden, C., Ackermann, F.: Cognitive mapping expert views for policy analysis in the public sector. Eur. J. Oper. Res. **152**, 615–630 (2004)
9. Ackermann, F., Eden, C.: Making Strategy: Mapping Out Strategic Success. SAGE, London (2011)
10. Garfinkel, H.: Ethnomethodology's program. Soc. Psychol. Q. **59**, 5–21 (1996)
11. White, L., Burger, K., Yearworth, M.: Understanding behaviour in problem structuring methods interventions with activity theory. Eur. J. Oper. Res. **249**, 983–1004 (2016)
12. White, L., Yearworth, M., Burger, K.: Understanding PSM interventions through sense-making and the mangle of practice lens. In: Kamiński, B., Kersten, Gregory, E., Szapiro, T. (eds.) GDN 2015. LNBIP, vol. 218, pp. 13–27. Springer, Heidelberg (2015). doi:10.1007/978-3-319-19515-5_2
13. Tavella, E., Franco, A.L.: Dynamics of group knowledge production in facilitated modelling workshops: an exploratory study. Group Decis. Negot. **24**, 451–475 (2015)
14. Franco, A.L., Greiffenhaggen, C.: Unpacking the complexity of group problem structuring. In: Kamiński, B., Kersten, G., Szufel, P., Jakubczyk, M., Wachowicz, T. (eds.) 15th International Conference on Group Decision and Negotiation Letters, pp. 33–34. Warsaw School of Economics Press, Warsaw (2015)
15. Callon, M.: Some elements of a sociology of translation: domestication of the scallops and the fishermen of St Brieuc Bay. In: Law, J. (ed.) A New Sociology of Knowledge?, pp. 196–229. Routledge, London (1986). Power, action & belief
16. White, L.: Understanding problem structuring methods interventions. Eur. J. Oper. Res. **199**, 823–833 (2009)
17. Hiltz, S.R., Dufner, D., Fjermestad, J., Kim, Y., Ocker, R., Rana, A., Turoff, M.: Distributed group support systems: theory development and experimentation. In: Olson, G.M., Malone, T.W., Smith, J.B. (eds.) Coordination Theory and Collaboration Technology. Psychology Press, UK (2013)
18. White, L., Taket, A.: The death of the expert. J. Oper. Res. Soc. **45**, 733–748 (1994)
19. Schein, E.H.: Process Consultation: Lessons for Managers and Consultants, vol. 2. Addison-Wesley, Boston (1987)
20. Tavella, E., Papadopoulos, T.: Expert and novice facilitated modelling: a case of a viable system model workshop in a local food network. J. Oper. Res. Soc. **66**, 247–264 (2015)

21. Scott, R.J., Cavana, R.Y., Cameron, D.: Recent evidence on the effectiveness of group model building. Eur. J. Oper. Res. **249**, 908–918 (2015)
22. Ventura, A., Dias, L., Clímaco, J.: On facilitating group decision making processes with VIP analysis. In: Zaraté, P., Kersten, G.E., Hernández, J.E. (eds.) GDN 2014. LNBIP, vol. 180, pp. 246–253. Springer, Heidelberg (2014). doi:10.1007/978-3-319-07179-4_28
23. Rouwette, E.A.J.A.: Facilitated modelling in strategy development: measuring the impact on communication, consensus and commitment. J. Oper. Res. Soc. **62**, 879–887 (2011)
24. Rouwette, E., Bastings, I., Blokker, H.: A comparison of group model building and strategic options development and analysis. Group Decis. Negot. **20**, 781–803 (2011)
25. Papamichail, K.N., Alves, G., French, S., Yang, J.B., Snowdon, R.: Facilitation practices in decision workshops. J. Oper. Res. Soc. **58**, 614–632 (2007)
26. Belton, V., Harvey, P., Rouwette, E., van Zijderveld, E.J.A., Morrill, N., Shaw, D.: Transferring PSM craft skills in a 2 hour session. In: 49th Annual Conference of the Operational Research Society, OR49, pp. 93–105 (2007)
27. Henning, P., Chen, W.C.: Systems thinking: common ground or untapped territory? Syst. Res. Behav. Sci. **29**, 470–483 (2012)
28. Buckle-Henning, P., Wilmshurst, J., Yearworth, M.: Understanding systems thinking: an agenda for applied research in industry. In: 56th Meeting of the International Society for the Systems Sciences, San Jose, USA (2012)
29. Keys, P.: OR as technology: some issues and implications. In: Keys, P. (ed.) Understanding the Process of Operational Research: Collected Readings. Wiley, Chichester (1995)
30. Keys, P.: On becoming expert in the use of problem structuring methods. J. Oper. Res. Soc. **57**, 822–829 (2006)
31. Keys, P.: OR as technology - some issues and implications. J. Oper. Res. Soc. **40**, 753–759 (1989)
32. Keys, P.: OR as technology revisited. J. Oper. Res. Soc. **49**, 99–108 (1998)
33. White, L., Burger, K., Yearworth, M.: Big data and behaviour in OR: towards a 'Smart OR'. In: Kunc, M., Malpass, J., White, L. (eds.) Behavioral Operational Research: Theory, Methodology and Practice. Palgrave Macmillan, Basingstoke (2016)
34. Yearworth, M., White, L., Ormerod, R.: The performative idiom and PSMs. In: 27th European Conference on Operational Research (EURO 2015), Glasgow, Scotland (2015)
35. White, L., Yearworth, M., Burger, K.: Understanding PSM interventions through sense-making and the mangle of practice lens. In: Kamiński, B., Kersten, G., Szapiro, T. (eds.) 15th International Conference on Group Decision and Negotiation (GDN 2015). Warsaw, Poland (2015)
36. Connell, N.A.D.: Evaluating soft OR: some reflections on an apparently 'unsuccessful' implementation using a Soft Systems Methodology (SSM) based approach. J. Oper. Res. Soc. **52**, 150–160 (2001)
37. Packer, M.J., Goicoechea, J.: Sociocultural and constructivist theories of learning: ontology, not just epistemology. Educ. Psychol. **35**, 227–241 (2000)
38. White, L.: Evaluating problem-structuring methods: developing an approach to show the value and effectiveness of PSMs. J. Oper. Res. Soc. **57**, 842–855 (2006)
39. Callon, M.: The role of lay people in the production and dissemination of scientific knowledge. Sci. Technol. Soc. **4**, 81–94 (1999)
40. White, L.: Behavioural operational research: towards a framework for understanding behaviour in OR interventions. Eur. J. Oper. Res. **249**, 827–841 (2016)
41. Midgley, G., Cavana, R.Y., Brocklesby, J., Foote, J.L., Wood, D.R.R., Ahuriri-Driscoll, A.: Towards a new framework for evaluating systemic problem structuring methods. Eur. J. Oper. Res. **229**, 143–154 (2013)

42. Davis, J., MacDonald, A., White, L.: Problem-structuring methods and project management: an example of stakeholder involvement using hierarchical process modelling methodology. J. Oper. Res. Soc. **61**, 893–904 (2010)
43. Checkland, P., Scholes, J.: Soft Systems Methodology in Action: Including a 30-Year Retrospective. Wiley, Chichester (1999)
44. Westcombe, M., Franco, L.A., Shaw, D.: Where next for PSMs - a grassroots revolution? J. Oper. Res. Soc. **57**, 776–778 (2006)
45. Ranyard, J.C., Fildes, R., Hu, T.-I.: Reassessing the scope of OR practice: the influences of problem structuring methods and the analytics movement. Eur. J. Oper. Res. **245**, 1–13 (2015)

# Negotiations

# Bargaining Power – Measuring it's Drivers and Consequences in Negotiations

Tilman Eichstädt[1] [ID], Ali Hotait[2]([⊠]) [ID], and Niklas Dahlen[3] [ID]

[1] bbw University of Applied Sciences,
Leibnizstrasse 11-13, 10625 Berlin, Germany
[2] University of Erfurt, Nordhäuser Str. 63, 99089 Erfurt, Germany
ali.hotait@uni-erfurt.de
[3] HHL Leipzig Graduate School of Management,
Jahnallee 59, 04109 Leipzig, Germany

**Abstract.** In this study, the authors carried out a laboratory experiment with professionals of purchasing departments and examined the effects of negotiation power on the outcome of distributive bargaining. The participants took over the role of buyers and sellers alternatively. Power was operationalized in terms of BATNA, time pressure, asymmetric information and self-constraint based on the different theoretical concepts of power applied in social sciences. The results support the influence of the individual factors as predicted by theory to a very large extent. Especially BATNA, time preferences and information differences have a great influence on negotiation outcomes. Hence, the main purpose of the present paper is to develop a basis for a more consistent operationalization of drivers of bargaining power and its influence on negotiation performance.

**Keywords:** Negotiation · Bargaining · Power · Time pressure · BATNA

## 1 Introduction

It is widely known that negotiation performance depends to a large extent on the distribution of bargaining power between negotiation parties. However, up to now there is no comprehensive understanding regarding the key elements of bargaining power and their interrelation, power is typically regarded as a vague concept [1]. In fact, given the variety of concepts of power in general and of bargaining power in particular [2] many researchers see an increasing need for operationalizing power in a more context-specific way [3].

In his seminal book, Howard Raiffa defines the "Best Alternative to a Negotiated Agreement" (BATNA) as a key element of every negotiation because negotiations will only lead to a result if the parties find an agreement between their perceived BATNAs. Generally, Raiffa shows that the better your BATNA the better your power position in the negotiation [4, 5]. We, as well as most other researchers, agree with this perspective to form the basis for the discussion. Beyond that, however, it is not well understood so far, what defines the exact outcome between the two BATNAs of the bargaining parties and which other elements of bargaining power help either party to claim a bigger share of the bargaining "pie". In his standard work on negotiation analysis, Raiffa describes

© Springer International Publishing AG 2017
D. Bajwa et al. (Eds.): GDN 2016, LNBIP 274, pp. 89–100, 2017.
DOI: 10.1007/978-3-319-52624-9_7

power as "a multifaceted concept", where advantageous alternatives, information or even skills, but also other factors play an important role [5]. We would like to develop a better understanding of which factors are key in driving one's party's bargaining power beyond improving your BATNA. In other fields of social research, there are several important scientists who identified relevant approaches to define power such as the psychologists French and Raven [6], sociologist Emerson [7] and later the economist Galbraith [8]. Regarding power and influence in negotiations, important contributions from an economic or game-theoretic perspective have been made by Nash [9], Schelling [10], and on behalf of Rubinstein [11]. The first attempt to connect the independent research streams, game-theoretic research and behavior-related research, was conducted by Howard Raiffa, who is considered the father of negotiation analysis by many experts in this field [12]. In this paper we follow Raiffa's footsteps and try to assess power as a multidimensional concept which can only be understood by combining the insights of various research streams.

The following research paper focuses on distributive, two-party negotiations as typically seen between seller and buyer, irrespective of the fact whether it is a single- or a multi-dimensional negotiation. In order to gain a better understanding of bargaining power, the paper begins with a short comparison of the existing definitions of power in social interactions and the identification of their possible impact on negotiation performance in a standard distributive two-party negotiation scenario. Based on this we will develop a synthesized concept of the key factors driving negotiation power and we will present some first empirical data to support our concept. Against this background, the main objective of the present paper is to understand how changing the bargaining power setting impacts the negotiators performance. In order to gain profound insights, we want to derive valuable implications for the buying and selling practice, as well as for the understanding of the concept of negotiation power in general.

## 2    Theoretical Background

Power has been defined by the sociologist Max Weber as: "*the possibility of imposing one's will upon the behavior of other persons*" [13, p. 324]. This definition fits into the typical distributive bargaining context, where the negotiators strive to maximize their individual utilities. Before focusing on a distributive negotiation context it needs to be clarified what is meant by "negotiation". Voeth and Herbst offer a comprehensive definition that describes negotiation as a process that entails following attributes: (1) Involvement of multiple parties; (2) Goal congruence; (3) Conflicting parties; (4) Zone of Possible Agreement (ZOPA); (5) Interactive process [14]. Bazerman and Neale speak of distributive negotiation when referring to negotiations about a single issue [15, p. 16]. This key characteristic of distributive negotiations is reflected by a statement of Walton and McKersie stating that distributive negotiation deals with an issue whereas integrative bargaining deals with problems [16]. A well-known example for distributive negotiation are price negotiations [15, p. 72]. The objective of a party involved in distributive negotiations is to claim a greater part of the negotiation outcome and consequently: "*One party's gain is the other party's loss.*" [15, p. 72]. In this situation the questions, who has how much bargaining power, and how can we evaluate the parties' bargaining power beyond the parties' BATNA. Even though the term was coined by Fisher and Ury [17], Bazerman and Neale offer a comprehensive definition: *"the lowest value acceptable to you for a negotiated agreement"* [15, p. 68]. The concept of BATNA is not uncontroversial because one's individual BATNA does not need to be congruent with the reservation price as other factors such as time pressure or relationships do not affect the party's BATNA but the reservation price [18]. Still, Sebenius argues that the BATNA plays a tactical role because it changes over time and the active search for alternative agreements can improve the BATNA and, thus, the negotiation situation [19]. In addition to this, it is true that BATNA can have a significant influence on power in negotiations [20]. Assuming that the initial situation for two negotiators is the same, the party with the better BATNA claims more of the subject [5].

There have been numerous approaches to define power. The most commonly used definition in social research has been introduced by French and Raven in 1959, when they originally identified five bases of social power: reward, coercion, legitimate, expert and referent [21]. This concept has been enhanced by Raven and other authors to a concept of six bases of power (including informational power) and further differentiating the individual components, for example into personal and impersonal reward and coercion. Based on the concept, series of articles evaluate the impact and importance of bases of power in differing contexts [6].

Although the term bargaining power is frequently used, only few others have tried to translate the French and Raven definition of power into the typical bargaining context. Lewicki et al. propose a concept of the following five sources of power: Informational, Personality, Position-based, Relationship-based and Contextual. In their concept, the term position-based power includes both legitimate power such as formal authority as well as resource-control as a basis for reward power [22]. Contextual power includes the concept of a BATNA as well as cultural definitions of power. In addition, further meaningful approaches to build a more holistic model of bargaining power have been made by Kim and Fragale [2] and Kim et al. [23]. Unfortunately, there has been little to no follow-up on their work, and no empirical testing of their models in a typical buyer-seller negotiation context.

Independently of the French and Raven approach the famous economist Kenneth Galbraith developed an alternative approach to define power by reducing power in social or economic interaction to only three instruments [8]: (1) Condign power (force based on the prospect of punishment); (2) Compensatory power (force based on the prospect of any reward or compensation); (3) Conditioned power (force in the belief that the effect will be virtuous or proper). In addition to these three instruments, he defines three sources of power: personality, property and organization. In daily life, power is typically enforced by combining sources of power with instruments of power (e.g. property with compensatory power or organization with conditioned power). Galbraith even states that never in the consideration of power is only one source or one instrument at work [8].

Whereas the Galbraith approach to power contains some straightforward parallels to French and Raven, e.g. reward and compensatory power and condign power and coercive power, a significantly different access to power has been given by Emerson and his power dependence theory [7] which was adopted by many other researchers, such as Bacharach and Lawler [24]. In the center of Emerson's concept is the idea that the power P of actor X over actor Y is equal to and based upon the dependence of actor Y upon X [7].

$$Pxy = Dyx. \tag{1}$$

$$Pyx = Dxy. \tag{2}$$

This is obviously an important approach to negotiation, which needs to be considered, since negotiations typically involve two parties which both have some sort of shared goal and hence, the power of both parties needs to be considered. Basically, Emerson already prepared the ground for the BATNA concept as he concluded that in order to rebalance a power relationship an important approach is to either cultivate or to deny the alternative sources that both sides have to achieve their personal goals [7]. This is basically the same approach given by most negotiation experts when recommending improving one's BATNA as an important step to improve one's bargaining position. However, considering Emerson's findings (or further elaborations such as those of Kim, Pinkley and Fragale) [23] in the specific context of buyer-seller negotiations, a specific challenge arises: Each party's dependence in a buyer-seller negotiation is in its center dependent on the difference between the final price agreed and their BATNA. Emerson defines the dependence of actor X upon actor Y as being directly proportional to X's motivational investment in goals mediated by Y. In a buyer-seller context with the focus on the price negotiation, the goals mediated by the negotiation are the individuals' shares of the surplus from the negotiation. These, however, depend on the final price agreement [7]. If the final price is very close to the BATNA of party X the surplus X gains from the agreement are very small. Hence, its dependence on a negotiated agreement should be small as well. Accordingly, the dependence becomes very large when the expected surplus from the negotiated agreement increases. This means that in basically every buyer-seller negotiation each party's negotiation power is naturally limited by the fact that the more power one side applies successfully to drive the price towards its preference, the less benefit and hence the less dependence of the other party remains. Basically, we see that Emerson's approach to power confirms what is considered common knowledge in negotiation literature today: The fact that the BATNA is a key driver of negotiation power.

*H1: The BATNA is a key element of negotiation power as negotiators with an improved BATNA will realize a larger share from the initial ZOPA.*

In addition, Emerson states that the other opportunity the parties have to rebalance the power relationship is to change their motivational investment into the goals mediated by the cooperation between party a and b [7, 24]. This is also considerably important for negotiators: The less you care about the agreement in a negotiation, the more power you have. First, you appear to have a high BATNA and second, you lose dependence on the other side and a negotiated agreement. In fact, this second line of thought appears to be consistent with the behavior of many negotiators to underrepresent their interest in an agreement and to act as if they would be well off without an agreement as well [25]. This finding is also consistent with the game theoretical perspective on bargaining games. In a Nash bargaining situation both parties have a strong incentive to display less gain or utility increase from the bargaining solution than they actually do have [26].

Taking this thought even further, it shows that Emerson's approach is also consistent with Nobel laureate Tom Schellings' most prominent finding on power in negotiations. He states that one's power in negotiations depends on one's ability to bind oneself on certain constraints ("the power to constrain an adversary rests in the power to bind yourself") [10]. Imposing oneself a credible amount of self-constraint is nothing more than a reduction in one's own motivational investment. Apparently, one side is willing to give up a negotiated solution as long as certain constraints are not fulfilled. In a buyer-seller negotiation context this constraint is often given as a certain prerequisite, e.g. each side tries to impose the other side's acceptance of its corporate terms and conditions as a binding constraint for any further negotiation. In ongoing negotiations, one side may show reduced motivational investment by threatening to break up negotiations if a certain price level is not set. Basically, this element of negotiation power is also consistent with French and Raven's definition of "Coercive Power" [21, p. 253] or Galbraith concept of "Condign Power" [8, p. 23] as far as these concepts can be applied in a buyer-seller situation where we assume that the biggest possible threat is to withdraw from the negotiation. Fortunately, the influence of coercive power and threats is also confirmed as an important part of bargaining games by game theorists. John Nash included threatening strategies into his classic bargaining model already in 1953 [25].

*H2: Coercive power is a key element of negotiation power. If one party can more credibly threaten to withdraw from a negotiated agreement under certain constraints, it will realize a larger share from the initial ZOPA.*

An important aspect of the French and Raven power concept, which is not reflected by Emerson and which does not appear either in Galbraith's anatomy of power, is the importance of information. In fact, informational power was not included in the very first concept of 1959, however, it was included in a later publication by Raven in 1965. Raven described informational power as the ability of an agent to bring about change through the resource of information [21]. Economists and game theorists have established an extensive toolset to analyze negotiations with complete information, however, they do struggle with the prominent situation of two-sided incomplete information. The Myerson-Satterthwaite theorem shows by the means of game theory that there is no efficient solution for a simple two-party distributive bargaining situation as long as each side has secret reservation values [27]. Incomplete information is seen by economists as a key source of inefficiency as it might cause delays in the negotiation or even a break-up in a situation where a positive zone of potential agreement exists [28, 29]. Solutions for bargaining games with complete information were developed by Nobel laureates Nash [9] and later Harsanyi [30]. However, most buyer-seller negotiations are taking place in a set-up with incomplete information, since typically both sides do not know the exact reservation value or BATNA of the other side. In a situation, where one side has more precise information about the other side's reservation value or BATNA, this side is clearly at a more advantageous position [31]. Too much of complex algebra is not required to reason that both Raven's concept of informational power and game theoretic results underline that information is an important element for defining negotiation power.

*H3: Informational power is a key element of negotiation power. If one party is better informed about the other party's reservation values or BATNAs than the other party, it will realize a larger share from the initial ZOPA.*

Within the game theoretic literature one of the most important models for non-cooperative bargaining games was developed by Rubinstein [11]. In his analysis he confirms on a very general level that in any distributive negotiation the party with the higher time preference is at disadvantage compared to another party with lower time preference [32]. Aspects of different time preferences are not found so far in social-psychological approaches to power as in those of French and Raven or Emerson. Nonetheless, the ability to wait plays an important role for negotiators, which has also been recognized by Raiffa [5]. Surprisingly, there is little empirical analysis on the interplay between power resulting from a BATNA and power resulting from different time preferences [23].

*H4: Time is a key element of negotiation power. If one party acts under less time pressure, relative to the time pressure of the other party, it will realize a larger share from the initial ZOPA.*

## 3   Empirical Study

### 3.1   Methodology

**Negotiation Setting.** As stated above, the paper strives to close the existing research gap concerning the impact of changes in the negotiation setting on the negotiation outcome. In the review *"Thinking Back on Where We're Going: A Methodological Assessment of Five Decades of Research in Negotiation Behavior"*, Mestdagh and Buelens showed that in the last 40 years negotiation research was mostly conducted with student populations (approx. 80%) and with only 5% of practicing managers and private sector employees [33]. Also Herbst and Schwarz pointed out that *"only 3 percent of empirical negotiation-related studies are based on the experience of practicing manager"* [34, p. 148]. For instance, in their review Eliashberg et al. demand more practitioner-researcher interaction to address the areas neglected so far [35]. To address this shortcoming the study at hand was conducted in the course of an executive education program. The training was designed in order to enable the participants to understand that only marginal changes in the negotiation setting may have a tremendous impact on the negotiation outcome. Over the period of two years, ten to twelve participants per program were invited to join the negotiation experiment. In each experiment the participants were randomly assigned to dyads of buyers and sellers, negotiating 4 rounds to test each of the hypothesis in one specific round with a different negotiation partner and changing roles, so that there is no learning effect or is minimized. In total 130 participants (37.7% female and 55.4% male, 6.9% have not specified) joined and the results of 65 dyads per hypothesis can be analysed.

The demographic characteristics, e.g. age, sex, nationality and educational background, were randomly mixed in the dyads.[1]

**Negotiation Task.** Over four negotiation rounds with varying settings, as described below, the participants had to negotiate in the role of buyer or seller. Before the one-to-one negotiation started, the participants were briefed by a lecturer about the task. Although the negotiators were instructed to maximize their benefits, an agreement was not required, which means that the negotiation could end without an agreement. Buyer and seller received information about their exit points/reservation values for the below mentioned subjects of negotiation. Still, within the exit points it was not a prerequisite that an agreement needed to be reached. The participants had to negotiate a three years' contract which was supposed to represent a typical negotiation situation between an automotive supplier and an automotive producer. Irrespective the fact that in real-world setting automotive producers often have higher bargaining power the negotiation setting was simplified assuming that seller and buyer have the same initial position.

**Negotiation Rounds.** The negotiation rounds were manipulated according to one individual power lever:

- In the first experiment (BATNA) the buyer's negotiating power was reinforced with an alternative offer, so that the buyer had a concrete second offer comprising the pricing, start of delivery and payment terms.
- In the second negotiation experiment (additional information) both negotiating parties received specific information on their counterpart that was valuable to the negotiation process. The information revealed to the buyers was more important since the suppliers' manufacturing cost calculation was given.
- In the third negotiation experiment (time pressure) the negotiating power of the participants was restricted due to time pressure. The time pressure was induced by the amount of time participants were given to reach an agreement and the opportunity to get points withdrawed by time limits exceeded. In this experiment the buyer was under higher time pressure than the seller. The buyer had 10 min to reach an agreement and the seller 14 min. The negotiation setting simplifies and assumes that there is no difference of negotiation power due to the relationship between automotive producer and automotive supplier.
- In the fourth negotiation experiment (self-constraints) the sales representative had received a letter from his boss to his customers, who threatened to shut down production if no agreement at a high price level would be reached.

Every negotiation experiment was supposed to take 15 min. Negotiators were not forced to reach an agreement after the 15 min. Prior to the experiment the set-up was

---

[1] As part of the executive education program the participants were between 20 and 50 years old. Mainly German participants, but a certain percentage had a foreign background (French, English, Persian, Sri Lanka, American, South African). With regard to the educational background the groups were also mixed starting from people with a formal job training and ending with people with a Master degree. In each experiment the participants negotiated with a different participant in every round. The assignment was completely random with regard to demographic characteristics.

tested with an open end scheduled. It showed, that given the simple set-up, 15 min were sufficient.

**Negotiation Performance.** A scoring system was implemented in order to assess an individual's performance. In each round a maximum of 10 points could be reached. The performance was measured as a percentage of the ZOPA that could be claimed by the negotiator. For instance, a buyer who is able to claim 100% of the ZOPA is rewarded with 10 points in this negotiation round. In order to increase the competitive behavior of the participants and their strive for claiming as much as possible, they were incentivized by selecting the best negotiator at the end and by chocolate coins that they would receive for every point achieved.

## 3.2    Results

In order to answer our four hypotheses, we conducted a variance analyses (ANOVA) to examine the effects of our manipulations on the negotiators' outcome. Dyads which did not reach an agreement were removed from these analyses.

Our results of the study are represented in the following table. Table 1. presents the total dyads, means of ZOPA realized in percent, standard deviations, F-statistics and p-values of the dyads.

**Table 1.** ANOVA score table

| Hypothesis | Manipulation | Total dyad | Mean buyer (SD) | seller (SD) | F-ratio | p-value |
|---|---|---|---|---|---|---|
| H1 | BATNA | 64 | 64.09 (29.51) | 34.94 (28.89) | 31.90 | <.001*** |
| H2 | Self-constraint | 61 | 47.22 (38.53) | 52.64 (38.03) | .61 | .437 |
| H3 | Information | 64 | 62.58 (29.89) | 37.20 (29.76) | 23.17 | <.001*** |
| H4 | Time-pressure | 65 | 34.54 (19.03) | 65.38 (18.99) | 85.57 | <.001*** |

*p < .05 **p < .01 ***p < .001

*Effects of Alternatives on the Bargaining Power and on the Negotiators' Performance.* H1 states that the BATNA is a key driver of negotiation power and negotiators with an augmented BATNA will realize a larger share of the initial ZOPA. The ANOVA revealed that the buyer's negotiating power was reinforced with an alternative offer and that the buyer was able to settle a beneficial agreement (Buyer: M = 64.09, SD = 29.51; Seller: M = 34.94, SD = 28.89; $F(1, 126) = 31.90$, $p < 0.001$). Thus, our hypothesis H1 is confirmed.

*Effects of Self-constraint on the Bargaining Power and on the Negotiators'*
*Performance.* In H2, we proposed that coercive power is a key element of negotiation power, therefore, if one party can threaten to withdraw from a negotiated agreement under certain constraints more credibly, he or she will realize a larger share of the initial ZOPA. The results in Table 1 do not support our Hypothesis H2. The findings indicated that there were no significant differences between the participants (M = 52.64, SD = 38.03) versus (M = 47.22, SD = 38.53), $F(1, 119) = 0.61$, $p < 0.437$.

*Effects of Additional Information on the Bargaining Power and on the Negotiators'*
*Performance.* H3 proposes that informational power is a key element of negotiation power and assumes that if a single party is better informed about the opposing party's reservation values or BATNA's than the other party about his or her, he or she will realize a larger share of the initial ZOPA. The ANOVA on this information index yielded a main effect of the participants' informational power and indicated that a buyer with additional information reached higher results (M = 62.58, SD = 29.89) than the seller (M = 37.20, SD = 29.76), $F(1, 126) = 23.17$, $p < 0.001$. H3 is thus supported.

*Effects of Time Pressure on the Bargaining Power and on the Negotiators'*
*Performance.* As expressed in H4, time is a key lever of negotiation power and negotiators with relatively low time constraints are able to claim a larger share of the ZOPA than negotiators under high time pressure. The negotiating power of the participants was restricted due to time pressure and the buyer who was under higher time pressure than the seller obtained lower results than his counterpart. The ANOVA showed the following results: (M = 65.38, SD = 18.99) versus (M = 34.54, SD = 19.03), $F(1, 128) = 85.57$, $p < .001$. Therefore, H4 is supported.

## 4   Discussion

*Findings.* In the research at hand we tried to close the gap between theoretical models and the practical implementation of negotiation by examining the levers of bargaining power, a key driver of negotiators' performance. It goes without saying that negotiation power plays a crucial role. However, to the best of our knowledge there is no comprehensive framework unifying the key levers on negotiation power. Our experiment displays strong evidence for the following levers to have a key influence on negotiators' performance: alternative offers, additional information and time pressure. Surprisingly, we could not find any significant influence on behaviour of coercive power, imposed by threatening potential.

*Limitations.* The multitude of negotiation models and various overlapping research streams indicate that there are various approaches to negotiation research. Consequently, when interpreting our findings one should bear in mind the following caveats: Firstly, the experiment took place in a simplified negotiation setting, assuming that the participants only negotiate three criteria and have limited information. Both parties were encouraged to improvise whenever they considered this to be appropriate. Moreover, asking people to self-assess their negotiation skills did not yield any significant results with regard to negotiation power. In addition to this, Backhaus et al.

demonstrated that buyer-seller negotiations are often team negotiations, which constitutes a deviation from our one-to-one negotiation setting [36]. In general, it would be very interesting to see if negotiation performance is influenced by the size of the negotiation team. Can a two-party-team realize better results on average than a one-person-team? Furthermore it is worthwhile to analyse how negotiation performance develops over several rounds with the same negotiation team and a certain relationship development.

A key question is to which extent the individual levers were manipulated in the respective negotiation rounds to influence bargaining power. With regard to alternatives, information and coercive power, we adjusted the influence to the same level of giving one party an advantage which should result in a 75:25 distribution of the ZOPA. With regard to time at hand the advantages were smaller. In future settings it should be assessed how time could be manipulated accordingly and which effect the length of a negotiation has on the result as well as the time pressure itself.

The negotiation outcome is a complex result of numerous influencing factors, which means that there are interdependencies between them. In our study we did not assume interdependencies. Finally, the experiment assumed ultimate distributive behaviour by the negotiators. In actual negotiations participants might take other actions as certain negotiations are recurring or impact the relationship between the parties.

The setting of the experiment did not exclude possible learning effects which might occur in the course of the program. However it excluded signalling effects as participants never negotiated with the same person twice or more often than once.

*Further Research.* The study revealed that certain levers might have a high impact on the negotiation outcome. Thus, further research should try to further operationalize the levers in an interdependent context and develop further approaches how to use them effectively. The study at hand did not shed light on the interaction of the different types of power. Further research should focus on identifying linkages between levers. It will also be important to understand more in detail how even small power differences influence the negotiation outcome, or to what extend minimum required power levels exist. In our study we removed dyads that did not reach an agreement from the analyses, because we do not know the reason for these no agreements or impasse rate. It would be possible that those happened actually due to threatening. Therefore further research should additionally pay attention on the reason for no agreements and may consider impasse rates as a dependent variable that should be treated like outcome and performance. If many dyads reach an impasse, then impasse rates themselves could be an interesting dependent variable. Impasse appears when negotiators do not reach an agreement. To the best of our knowledge, only a few studies have considered impasse rates as a dependent variable [37, 38]. Trip and Sondak claim that impasse rates have been mostly ignored as a dependent variable and that their absence may bias experimental results [39].

# References

1. Alavoine, C., Kaplanseren, F., Teulon, F.: Teaching (and learning) negotiation: is there still room for innovation? Int. J. Manag. Inf. Syst. **18**(1), 36–40 (2014)
2. Fragale, A., Kim, P.: Choosing the path to choosing the path to bargaining power: an empirical comparison of BATNAs and contributions in negotiation. J. Appl. Psychol. **90**(1), 373–381 (2005)
3. Krause, D., Kearney, E.: The use of power bases in different contexts: arguments for a context-specific perspective. In: Schriesheim, L.L.A. (ed.) Power and Influence in Organizations: New Empirical and Theoretical Perspectives. Information Age Publishing, Inc., Hartford (2006)
4. Raiffa, H.: The Art and Science of Negotiation. Belknap Press, Cambridge (1982)
5. Raiffa, H., Richardson, J., Metcalfe, D.: Negotiation Analysis: The Science and Art of Collaborative Decision Making. Harvard University Press, Cambridge (2007)
6. Raven, B., Schwarzwald, J., Koslowsky, M.: Conceptualizing and measuring a power/interaction model of interpersonal influence. J. Appl. Soc. Psychol. **28**(4), 307–332 (1998)
7. Emerson, R.: Power-dependence relations. Am. Soc. Rev. **27**(1), 31–41 (1962)
8. Galbraith, K.: The Anatomy of Power. Houghton Mifflin, Boston (1983)
9. Nash, J.: The Bargaining Problem. Econometrica **18**(2), 155–162 (1950)
10. Schelling, T.: The Strategy of Conflict. Harvard University Press, Cambridge (1960)
11. Rubinstein, A.: Perfect equilibrium in a bargaining model. Econometrica **50**(1), 97–110 (1982)
12. Sebenius, J.: Negotiation analysis: from games to inferences to decisions to deals. Negot. J. **25**(4), 449–465 (2009)
13. Weber, M.: Max Weber on Law in Economy and Society. Harvard University Press, Cambridge (1954)
14. Voeth, M., Herbst, U.: Verhandlungsmanagement: Planung, Steuerung und Analyse. Schäffer-Poeschel Verlag, Stuttgart (2009)
15. Bazerman, M., Neale, M.: Negotiating Rationally. Free Press, New York (1992)
16. Walton, R., McKersie, R.: A Behavioral Theory of Labor Negotiations: An Analysis of a Social Interaction System. McGraw-Hill, New York (1965)
17. Fisher, R., Ury, W.: Getting to Yes. Random House Business Books, New York (1992)
18. Wheeler, M.: Negotiation Analysis: An Introduction. Harvard Business School. Harvard Press, Boston (2002)
19. Sebenius, J.: Negotiation analysis: a characterization and review. Manag. Sci. **38**(1), 18–38 (1992)
20. Alfredson, T., Cungu, A.: Negotiation theory and practice: a review of the literature. FAO (2008)
21. French, J., Raven, B.: The bases of social power. In: Cartwright, D. (ed.) Studies in Social Power, pp. 150–167 (1959)
22. Lewicki, R., Saunders, D., Barry, B.: Negotiation. McGraw-Hill Irwin, Boston (2006)
23. Kim, P., Pinkley, R., Fragale, A.: Power dynamics in negotiation. Acad. Manag. Rev. **30**(4), 799–822 (2005)
24. Bacharach, S., Lawler, E.: Power and tactics in bargaining. Ind. Labor Relat. Rev. **34**(2), 219–233 (1981)
25. Nash, J.: Two-person cooperative games. Econometrica **21**(1), 128–140 (1953)
26. Binmore, K., Dasgupta, P.: Economic Organizations as Games. Blackwell, Oxford (1986)

27. Myerson, R., Satterthwaire, M.: Efficient mechanisms for bilateral trading. J. Econ. Theory **29**(2), 265–281 (1983)
28. Bester, H.: Non-cooperative bargaining and imperfect competition: a survey. Zeitschrift für Wirtschafts- und Sozialwissenschaften **109**(1), 265–286 (1989)
29. Muthoo, A.: A non-technical introduction to bargaining theory. World Econ. **1**(2), 145–166 (2000)
30. Harsanyi, J.: Approaches to the bargaining problem before and after the theory of games. Econometrica **24**(2), 144–157 (1956)
31. Cramton, P., Ausubel, L., Deneckere, R.: Bargaining with incomplete information. In: Hart, R.J. (ed.) Handbook of Game Theory. Elsevier, Amsterdam (2002)
32. Osborne, M., Rubinstein, A.: Bargaining and Markets. Academic Press, Inc., Cambridge (1990)
33. Mestdagh, S., Buelens, M.: Thinking back on where we're going: a methodological assessment of five decades of research in negotiation behavior. In: International Association of Conflict Management Conference, Melbourne (2003)
34. Herbst, U., Schwarz, S.: How valid is negotiation research based on student sample groups? New insights into a long-standing controversy. Negot. J. **7**(2), 147–170 (2011)
35. Eliashberg, J., Lilien, G., Kim, N.: Marketing negotiations: theory, practice and research needs. ISBM Report (1994)
36. Backhaus, K., van Doorn, J., Wilken, R.: The impact of team characteristics on the course and outcome of intergroup price negotiations. J. Bus.-to-Bus. Mark. **15**(4), 365–393 (2008)
37. Neale, M.: The effects of negotiation an arbitration cost salience on bargainer behavior: the role of the arbitrator and constituency on negotiator judgment. Organ. Behav. Hum. Perform. **34**(1), 97–111 (1984)
38. Malouf, M., Roth, A.: Disagreement in bargaining: an experimental study. J. Confl. Resolut. **25**(1), 329–348 (1981)
39. Tripp, T., Sondak, H.: An evaluation of dependent variables in experimental negotiation studies: impasse rates and pareto efficiency. Organ. Behav. Hum. Decis. Process. **51**(1), 273–295 (1992)

# A Deviation Index Proposal to Evaluate Group Decision Making Based on Equilibrium Solutions

Alexandre Bevilacqua Leoneti$^{(\boxtimes)}$ and Fernanda de Sessa

Ribeirão Preto School of Economics, Administration and Accounting,
University of São Paulo, São Paulo, Brazil
ableoneti@usp.br

**Abstract.** The equilibrium proposed by Nash provides a basis from which group decisions can be selected. This kind of choice establishes a situation in which none of the participants will have any incentive to change their strategy if they are acting rationally, which is the major assumption of game theory. Leoneti proposed a utility function that allows multi-criteria problems to be modeled as games in order to find alternatives that meet the Nash equilibrium conditions for solving conflicts in group decision-making process. The objective of this research was to propose a deviation index from the theoretical rational decision (the Nash equilibrium solution) and to discuss the use of this index as an indicator of the theoretical rationality deviation. In accordance with other results presented in the literature, it was found that the group might not always choose this alternative, deviating from the equilibrium solutions, measured here by a deviation index.

**Keywords:** Nash equilibrium · Game theory · Group-decision making

## 1 Introduction

In a negotiation individuals interact their own decisions and perceptions with those of others, considering the actions and reactions of all involved. This kind of situation is known as strategic interaction, since the results of a decision-maker do not depend only on their strategies, but also on the strategies of other decision makers. In this context, game theory is presented as a mathematical approach to model situations of group decision-making, considering the individual strategic actions of decision-makers to develop solutions that are more acceptable to the participants involved. On the other hand, game theory requires a major assumption about the players, which is the fact that they are rational [1, 2].

Assuming the hypothesis of rationality of players, one important step is the proposal of a solution for the games. Among the methods to solve a game, the best known is the equilibrium proposed by Nash [3]. In Nash equilibrium, players have assumptions about the strategies of their competitors and choose the best possible strategy, taking into account the possible choices of all other players. Other players, when acting equally, will lead to a situation in which none of the participants will have any incentive to change their strategy (if they are acting rationally).

© Springer International Publishing AG 2017
D. Bajwa et al. (Eds.): GDN 2016, LNBIP 274, pp. 101–112, 2017.
DOI: 10.1007/978-3-319-52624-9_8

According to Osborne and Rubinstein [1], when an equilibrium found is the best possible response of each rational player to the strategies of others, it is considered that the Nash equilibrium was found for the game. Epistemologically, according to Aumann and Brandenburger [4], "since each player knows the choices of the others, and is rational, his choice must be optimal given theirs; so by definition, we are at a Nash equilibrium". Therefore, the Nash equilibrium concept proposes a solution to the strategic behavior of the decision makers interacting in group decision-making, identifying strategies that increase the chances of create a contract between those involved, which can assist in resolving conflicts among players [5, 6].

In this context, Leoneti [7] proposed a utility function that allows the modeling of group multi-criteria problems as games in order to find alternatives that meet the Nash equilibrium conditions for solving conflicts in the group decision-making process.

However, people might occasionally make choices that are not consistent with the logic of rational choice. In this sense, Myerson [8] state that numerous experimental studies regularly find inconsistent behavior that violates the principles of rationality. Roth and Murnighan [9] evaluated indices for measuring the difficulty of achieving a cooperative equilibrium in repeated games. Gilboa and Matsui [10] affirm that even with players being rational there is no compelling reason to believe that they choose just equilibria solutions. According to Camerer and Fehr [11], aggregate outcomes in strategic games would be far from Nash equilibrium if the strategies are complementary. Bernheim [12] proposes to think the concept of rationality in terms of internal consistency and, under this assumption, it will not necessarily be deductive logic to expect players selecting Nash strategies all the time. Finally, Kahneman and Tversky [13] show that the classic normative analysis given by the Expected Utility Theory, proposed by von Neumann and Morgenstern [14], which is related to how individuals should behave (considering that they are rational), might not always be true.

Therefore, the objective of this research is to propose, among the Nash equilibria found from the application of the Leoneti method [7] in a group context, a deviation index from the theoretical rational decision (the Nash equilibrium solution) and to discuss the use of this index as an indicator of the theoretical rationality deviation. The index shows the proportion in which the group has deviated from the rational solution that, according to the theory, would create a stable solution to those involved in the decision. For proposing this measure, two movements were required: (i) the selection of one Nash equilibrium, when more than one is found; and (ii) the proposition of a deviation index of this selected Nash equilibrium solution.[1]

In the experiment presented in this research the equilibrium selected was always the one that presents the highest payoffs average among players involved in the decision. This choice was made in accordance with the inequality-aversion theory, which states that players prefer allocations that make high and more equals outcomes [15]. In other words, it was established that the Nash equilibrium with the highest average payoff among decision-makers would be the selected equilibrium and this equilibrium would be, therefore, the equilibrium to be compared with the solution reached by the group.

---

[1] It is important to stress that this index can be calculated using any equilibrium as reference, which may vary depending on the research purposes.

In accordance with other results presented in the literature, it was found that the group might not always choose this alternative, deviating from the equilibrium solution. This deviance was measured here by a deviation index.

# 2   Method

The first stage of the research was the development of an experiment to evaluate possible deviation of the Nash equilibrium solution that would be found based on the payoff tables created by the application of the Leoneti [7] method. In this sense, it was necessary to determine a decision problem, contextualize it, create a performance matrix[2] and define how and where the application would occur.

## 2.1   The Experiment: Decision Problem, Performance Matrix and Application

The decision problem concerns the choice of a travel destination to be held in a group, as follows: "In order to attract and retain customers, a travel company held a promotion. A group of people was randomly selected to travel with all expenses paid by the company. The promotion conditions are: the winners must travel together and the limit value of the trip is 5,000 reais each. The company will allocate the 5,000 reais only to the trip expenses and if any money remains it will be returned to the company. Congratulations, you are one of the winners selected! Whereas every winner has 10 days of vacation, you should negotiate with the other winners to decide the final destination." In order to provide information regarding five possible travel destinations, the authors of the present research created a performance matrix.

To present alternatives closer to reality, the matrix data was based on five real travel packages offered by Brazilian travel companies. This source of information was used to include into the performance matrix criteria such as hotel evaluation and overall cost. That information was also used to arbitrate, by the present researches, each destination's grades for shopping facilities, cultural attraction, landscape availability and safety, which were described not structurally in the respective travel packages. A Likert-type scale from 1 (low) to 9 (high) was used to score those latter criteria, while the previous were scored using alternative performance for each criterion. The performance matrix can be seen in Fig. 1, where: (i) Hotel evaluation (grade 1–5): the hotel evaluation on booking sites; (ii) Travel time (in hours): hours of trip to the destination; (iii) Length of stay (number of nights): number of nights included; (iv) Cost (in R$): cost of the package that includes accommodation, breakfast and airfare; (v) Shopping (grade 1–9): whether the destination is good for shopping; (vi) Cultural Attractions (grade 1–9): possible presence of museums, theaters, etc.; (vii) Nature (grade 1–9): possible presence of natural landscapes; (viii) Safety (grade 1–9): whether it is safe in terms of health conditions, violence and terrorism.

---

[2] The term "performance matrix" was adopted over other possible classifications to the criteria versus alternative matrix, such as, i.e., "decision matrix", "consequence matrix", etc.

| Alternatives | Criteria | | | | | | | |
|---|---|---|---|---|---|---|---|---|
| | Hotel evaluation | Travel time (hours) | Length of stay (nights) | Cost (R$) | Shopping | Cultural attractions | Nature | Safety |
| A: Punta del Este | 5 | 2.5 | 4 | 2,839 | 5 | 3 | 9 | 8 |
| B: New York | 3.5 | 12 | 6 | 3,700 | 9 | 7 | 3 | 6 |
| C: Santiago | 2.5 | 4 | 5 | 2,683 | 4 | 5 | 7 | 7.5 |
| D: Paris | 3 | 13 | 7 | 4,150 | 6 | 9 | 6 | 7 |
| E: Istanbul | 4 | 18 | 9 | 4,500 | 3 | 8 | 5 | 4 |

**Fig. 1.** Performance matrix

After the performance matrix creation, a group of five undergraduate students from the Business Management course of the Ribeirão Preto School of Economics, Administration and Accounting at the University of São Paulo participated in one application. The application occurred in a meeting room previously prepared with overhead projector and conference table, providing appropriate conditions for negotiation. The application took place on March 17th, 2015 and began with the presentation of the decision problem.

The application had two phases: (i) individual and (ii) group phase. In the individual phase, each participant received a printed form containing the performance matrix and a briefing including the instructions and rules of the application. Each participant ranked the alternatives A, B, C, D and E of the performance matrix and wrote down the result in the form as their initial ranking. The participants also ranked the eight criteria from the performance matrix in order to elicit their preference using the Ranking Order Centroid – ROC [16] – method, because its ease of use and interpretation for application involving groups [17].

During the group phase, a first negotiation round (settled to 15 min) occurred without the support of any decision aid method. The participants were informed that they should convince the group to agree with their better-ranked alternative (ideally the first one). Then, the participants negotiated about the possible travel destination and tried to reach a consensus, each one considering their initial ranking. In the meantime, the present researchers calculated the solutions based on the utility function proposed by Leoneti [7] using the weight vectors generated by the application of ROC. In the individual level, for means of testing the consistency of the initial ranking, the ELECTRE III method was chosen due to its similarity to the equilibrium method proposed by Leoneti [7], since both use distance measurements to rank alternatives and share the principles of pairwise comparison.

After the first negotiation round, the results calculated by the methods were shown to the participants in order to either evaluate or support the group decision, when a solution had not been reached.

## 2.2  Leoneti [7] Method and the Decision Game

The decision game proposed here is a non-cooperative[3] strategic game defined by the tuple $<N, A, C, \succsim_i>$, where N is the set of $n$ players (decision makers), A is the set of $m$ actions (alternatives), C is the set of $c$ benefit criteria, and $\succsim_i$ is the preference set over A for each player $i \in N$. The numeric representation of the set of preferences $\succsim_i$ jointly is a function $\pi: \Re_+^{c \times n} \rightarrow [0, 1]$, proposed by Leoneti [7]. This utility function shows the pay-offs for a decision game among decision makers that has three strategies: (I) maintain the initial choice; (II) choose the alternative proposed by an opponent; and (III) choose a different alternative from the alternatives proposed by an opponent. Equation 1 shows the utility function for the game with two players.

$$\pi(x, y) = \varphi(x, IA).\varphi(x, y).\varphi(y, IA) \tag{1}$$

where, $x$ is the initial alternative, $y$ is the alternative proposed by the opponent, $IA$ is the ideal alternative (the alternative composed with the maximum absolute values of each criteria[4]), $\varphi(x, IA)$, $\varphi(y, IA)$ and $\varphi(x, y)$ are given by the pairwise comparison function $\varphi: \Re_+^c \rightarrow [0, 1]$, according to Eq. 2.

$$\varphi(x, y) = \left[\frac{\alpha_{xy}}{\|y\|}\right]^\delta . \cos\theta_{xy}, \text{ where } \delta = \begin{cases} 1, & \text{if } \alpha_{xy} \leq \|y\| \\ -1, & \text{otherwise} \end{cases} \tag{2}$$

where, $\alpha_{xy} = \|x\| \cos\theta_{xy}$ is the scalar projection of the vector $x$ on the vector $y$, $\cos\theta_{xy}$ is the angle between the two vectors, $\|y\| = \sqrt{y_1^2 + y_2^2 + \ldots + y_c^2}$ is the norm of the respective vector. The image of $\varphi$ (range of the function values) varies between 0 and 1 (due to the conditional $\delta$), meaning the closer it is to 1 the more similar are the alternatives. The joint utility function for games where the number of players is more than two is given by Eq. 3.

$$\pi(x, Y) = \varphi(x, IA). \prod_{i=1}^{n-1} \varphi(x, y_i).\varphi(y_i, IA) \tag{3}$$

where $n$ is the number of players, and $\pi(x, Y)$ defines, for a determined player, the payoff for all strategies (I, II or III) for an alternative $x$ when trading it with another set of alternatives $Y(y_i)$ proposed by all other players. The use of the joint utility function generates the payoff tables for all players, which estimates a utility measure for every possible strategy in the set of actions. Mathematically, if one of the terms (pairwise comparison function) of the utility function is close to zero (low similarity between any pair of alternative), then $\pi(x, y_i)$ tend to zero, which means that only similar alternatives

---

[3] The game is considered a non-cooperative game since the participants cannot make binding agreements before choosing their alternatives.

[4] This alternative is called "ideal" because it contains the maximum absolute value of all criteria considered in the alternative's evaluation and, therefore, is used as an indicator of direction to the maximum value that each criteria can eventually reach.

closed to IA are going to be considered in what is called "kernel" of the game. Therefore, a distinction between the preferable trades will be possible and for this reason, the matrices composed by $\varphi(x, y)$ are called trade-off matrices with the feature of being asymmetric. The players' likely strategies consider the fact that players might trade for alternatives that have high values given by the pairwise comparison function between the alternatives and the ideal solution. In other words, it is derived from the fact that they want to increase or at least keep their outcome in a trade.

Applying the function $\pi(x, Y)$ to each weighted performance matrix, which are weighted using the weight vector generated by the ROC method for each participant, will generate the payoffs table for all possible sets of strategies of the game (I, II or III) for each player. These payoff tables are the framework of the game translated from the original multicriterial approach. Figure 2 presents the framework for a game with two players, two alternatives and C criteria, where <2, [A, B], C, $\pi_i$>, from which Nash equilibria can be calculated.

|   | A | B |
|---|---|---|
| A | $\pi_1(A,A)$ | $\pi_1(A,B)$ |
| B | $\pi_1(B,A)$ | $\pi_1(B,B)$ |

|   | A | B |
|---|---|---|
| A | $\pi_2(A,A)$ | $\pi_2(A,B)$ |
| B | $\pi_2(B,A)$ | $\pi_2(B,B)$ |

**Fig. 2.** Framework for a game with two players and two alternatives

Figure 3 presents a scenario of possible strategic interactions between the players for the game presented in Fig. 2. (It is noteworthy that in a game with 2 alternatives there will not exist strategy III).

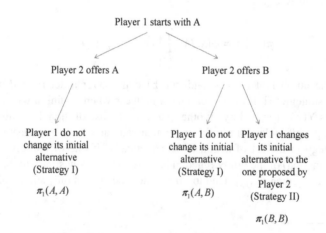

**Fig. 3.** Example of possible strategic interactions

### 2.3   Calculating, Disclosing and Evaluating the Results

In order to calculate the results from the Leoneti [7] method, GAMBIT [18] software was used to identify all pure Nash equilibria from the individual payoffs table. Based on the theory of inequality-aversion [15], a Nash equilibrium was then selected based on the players' highest average payoffs among the equilibria found. The results from ELECTRE III were obtained using SANNA [19] software, which calculated the individual ranking using the correspondent weight vector generated by ROC method. In order to calculate the results from ELECTRE III the criteria were all considered as true-criteria (without pseudo-criteria), hence the parameters veto, indifference, and preference were set to zero.

In sequence, the results were presented. If a consensus had already been reached by the group, each participant individually would respond to two questions available in their respective printed form. Both questions, namely: (i) do you feel satisfied with the group decision?; and, (ii) do you believe that the group decision was fair?, used a Likert-type scale from 1 (strongly disagree) to 7 (strongly agree) to evaluate the level of satisfaction and justice sense of each individual with the alternative chosen by the group. If a consensus had not been reached, the participants would compare their initial ranking with the results from the ELECTRE III method and, then, evaluate the results of the group from the Leoneti [7] method, for a second round of negotiation (settled to 5 min). If appropriate, those two question would be responded by the end of the second round of negotiation.

Finally, since the objective of the research was to evaluate the solution deviation from a specific Nash equilibrium solution based on the application of the method proposed by Leoneti [7], the index proposed in this research was calculated and the group decision was evaluated based on the comparison of the individual satisfaction and justice sense level with the index score.

Figure 4 summarizes the method including the decision problem, the performance matrix, the application of the Leoneti [7] method and the decision game.

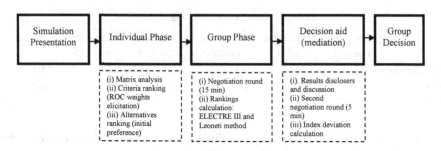

**Fig. 4.** Schematic presentation of the method

## 3   Results and Discussion

In this application, after the first round of negotiation, it was found that the group did not choose the selected Nash equilibrium solution (alternative D) as the solution to the problem. Although most decision makers defended alternative D, a decision maker

(Decision Maker 1) firmly opposed it, leading the group to choose alternative C in the second round of negotiations (Table 1 shows the initial ranking and the ranking calculated by the ELECTRE III for all participants). Considering the evaluation of justice sense and satisfaction, it was observed that Decision Maker 1 showed high satisfaction and sense of justice regarding the group decision (Table 2). Likewise, Decision Maker 3, which initially had chosen the alternative E as preferred, considered the group decision a fair decision because, in his opinion, "by the end all players had agreed with one alternative". However, his satisfaction was low, which could be explained because alternative C was his second worst alternative in his initial ranking (although ELECTRE III indicated the opposite). Similarly, the level of satisfaction and justice sense was not equal for the Decision Maker 2. This participant evaluated the decision as not fair because, in his opinion, "the group chose to cooperate more with who strongly denied to make concessions (Decision maker 1)".

**Table 1.** Results (individual level).

|  |  | $1^a$ | $2^a$ | $3^a$ | $4^a$ | $5^a$ |
|---|---|---|---|---|---|---|
| Decision maker 1 | Initial ranking | C | E | A | B | D |
|  | ELECTRE III | D | E | C | B | A |
| Decision maker 2 | Initial ranking | B | D | A | E | C |
|  | ELECTRE III | D | C | E | B | A |
| Decision maker 3 | Initial ranking | E | D | B | C | A |
|  | ELECTRE III | C | D | E | B | A |
| Decision maker 4 | Initial ranking | B | D | C | A | E |
|  | ELECTRE III | C | D | A | E | B |
| Decision maker 5 | Initial ranking | A | C | D | B | E |
|  | ELECTRE III | D | A | C | B | E |

**Table 2.** Satisfaction and justice sense.

|  | Decision maker 1 | Decision maker 2 | Decision maker 3 | Decision maker 4 | Decision maker 5 | Average |
|---|---|---|---|---|---|---|
| Satisfaction | 7 | 5 | 3 | 6 | 6 | 5.4 |
| Justice sense | 6 | 2 | 6 | 6 | 7 | 5.4 |

Considering just the first alternative in individual rankings, the difference between the initial ranking and that proposed by the ELECTRE III method was more noticeable for the Decision Maker 1, who revealed that have already been living in Paris (Alternative D of the performance matrix). This information might explain the behavior of this player that attributed to the alternative D several other factors and personal impressions of the city, which did not appear in the performance matrix. Thus, the alternative supported by him in the negotiation (Alternative C) did not correspond necessarily to his evaluation of criteria importance. Controversially, the calculation of ELECTRE III suggested that the ideal alternative to the Decision Maker 1 would be alternative D.

For evaluating the group decision (Table 2), it is recalled that the selected Nash equilibrium calculated from the payoffs generated by the utility function [7] indicated alternative D as the alternative with the highest payoffs for group consensus as can be seen in Table 3. If the Decision Maker 1 had acted more in favor of group's members, it can be said that the alternative D would also have been the final choice of the group, not having any deviation from this equilibrium. However, the group chose alternative C, creating a deviation from the highest payoff average based Nash equilibrium solution, which, according to the inequality-aversion theory [15], might explain the lower average among the scores given to satisfaction and justice sense.

**Table 3.** Results (group level).

|  | 1° | 2° | 3° | 4° | 5° |
|---|---|---|---|---|---|
| Pure Nash equilibria found[a] | D | E | C | B | - |
| Alternative chosen by the group | C | | | | |

[a]It was found four pure Nash equilibria calculated from the application of the utility function proposed by Leoneti [7]. The Nash equilibria found were ranked based on the criteria of high averages among payoffs. The alternative A did not belong to any Nash equilibria found

### 3.1 Deviation Index

The deviation index of the Nash equilibrium solution with highest average payoffs shows the proportion of how the group has deviated from this particular solution with highest average payoff among equilibria. It is proposed here that when this index is zero, it means that the group choose the equilibrium solution with the highest average payoff among players. When the index is different from zero, this may be negative or positive. A positive deviation index indicates that the choice of the group is a solution with a higher average payoff than the Nash equilibrium and, therefore, more interesting to the group decision (a Pareto improvement). A negative deviation index indicates that the average payoff of the decision made by the group is lower than the Nash equilibrium, so the group would gain by changing the choice for the Nash equilibrium. Equation 4 presents the deviation index proposed in this research.

$$\Delta_e = \frac{\bar{x}_g}{\bar{x}_e} - 1 \qquad (4)$$

where $\Delta_e$ is the deviation index of the selected Nash equilibrium solution, $\bar{x}_g$ is the average payoff of the group choice and $\bar{x}_e$ is the highest average payoff among all Nash equilibria found (selected Nash equilibrium). If $\bar{x}_g$ is larger than $\bar{x}_e$ the ratio between these two numbers is greater than 1 and therefore $\Delta_e$ will be positive. In this case, there was a deviation from the selected Nash equilibrium (with the highest average payoffs in this case), but the group choose a solution that provides, on average, higher payoffs to those involved. On the other hand, if $\bar{x}_g$ is lower than $\bar{x}_e$, $\Delta_e$ will be negative, which means that the group choose a solution that provides, on average, lower payoffs than the selected Nash equilibrium.

In this application, the deviation index was a negative 0.332 (Table 4), which interprets the fact that the group decision had average payoffs below of that provided by the Nash equilibrium with highest average payoffs. Consequently, it should not be expected that this choice would indeed satisfy the group as a whole. In fact, the analysis of this application shows that one of the participants (Decision Maker 1) held subjective analysis of criteria (based on past experiences) and did not act in favor of a group consensus. His behavior led the group's decision to the alternative C, with belongs to a Nash equilibrium with the third lower average payoffs (Table 3).

**Table 4.** Deviation index of the group choice from the selected Nash equilibrium solution.

| | Decision maker 1 payoff | Decision maker 2 payoff | Decision maker 3 payoff | Decision maker 4 payoff | Decision maker 5 payoff | Average payoff | $\Delta_e$ |
|---|---|---|---|---|---|---|---|
| Selected Nash equilibrium (alternative D) | 0.396 | 0.507 | 0.254 | 0.179 | 0.185 | 0.304 | −0.332 |
| Group choice (alternative C) | 0.103 | 0.081 | 0.106 | 0.245 | 0.482 | 0.203 | |

People often make choices that are not consistent with the logic of rational choice [13, 20]. Johnson-Laird and Shafir [20] stated that people are not intuitively logical, intuitively statistical, or decision theoretical intuitively rational. The accuracy of their thoughts and decisions would be the result of a complex and unobservable mental process. Milikkovic [21] further states that the perfect rationality might be only theoretical, even though the theory of rational choice accepts it as truth. This may be related to the existence of deviations and the fact that every negotiation varies, involving different rationality of people and levels. In this application, this deviation was measured by the negative value of the deviation index of the Nash equilibrium solution with highest average payoff.

## 4    Conclusions

This article provides the proposition of a deviation index that reveals the proportion in which the group deviates from a specific equilibrium solution that, according to rational principles, would be the most satisfying to those involved in the decision.

Based on the statement that players prefer more equal allocations of high outcomes, an experiment tested the Nash equilibrium with the highest average payoffs as the rational decision of a group. However, it was found that, corroborating results in the literature, groups do not always choose this alternative, deviating from the ideal Nash equilibrium solution. This deviation was measured here based on the relation between the choice made by the group and the equilibrium solution.

The proposal of the deviation index might help the intervention process, leading the decision to a more favorable outcome by putting the group aware of lower agreements made. Therefore, when the assessment of the criteria is necessary and there is difference of opinion in the group, the Leoneti [7] method and the deviation index proposed in this research can be used for supporting the decision making process toward the search of solutions considering the concepts of rationality.

**Acknowledgments.** The authors thank the National Council of Technological and Scientific Development (CNPq) for Regular Research Grant (458511/2014-5), and the São Paulo Research Foundation (FAPESP) for the Scientific Initiation Scholarship (2014/09540-0) and for the grant for Paper Presentation (2016/03722-5). The authors also acknowledge the helpful comments of two anonymous referees.

# References

1. Osborne, M.J., Rubinstein, A.: A Course in Game Theory. The MIT Press, Cambridge (1994)
2. Myerson, R.B.: Game Theory Analysis of Conflict. First Harvard University Press paperback edition (1997)
3. Nash, J.: Non-cooperative games. Ann. Math. **54**, 286–295 (1951)
4. Aumann, R., Brandenburger, A.: Epistemic conditions for Nash equilibrium. Econom.: J. Econom. Soc. **63**(5), 1161–1180 (1995)
5. Lee, C.: Multi-objective game theory models for conflict analysis in reservoir watershed management. Chemosphere **87**(6), 608–613 (2012)
6. Leoneti, A.B., Oliveira, S.V.W.B., Oliveira, M.M.B.: The Nash equilibrium as a solution to the conflict between efficiency and cost in the choice of systems for sanitary sewage treatment using a decision making model. Engenharia Sanitária e Ambiental **15**, 53–64 (2010). (in Portuguese)
7. Leoneti, A.B.: Utility function for modeling group multicriteria decision making problems as games. Oper. Res. Perspect. **3**, 21–26 (2016)
8. Myerson, R.B.: Nash equilibrium and the history of economic theory. J. Econ. Lit. **37**(3), 1067–1082 (1999)
9. Roth, A.E., Murnighan, J.K.: Equilibrium behavior and repeated play of the prisoner's dilemma. J. Math. Psychol. **17**(2), 189–198 (1978)
10. Gilboa, I., Matsui, A.: Social stability and equilibrium. Econom.: J. Econom. Soc. **59**(3), 859–867 (1991)
11. Camerer, C.F., Fehr, E.: When does "economic man" dominate social behavior? Science **311** (5757), 47–52 (2006)
12. Bernheim, B.D.: Rationalizable strategic behavior. Econom.: J. Econom. Soc. **52**(4), 1007–1028 (1984)
13. Kahneman, D., Tversky, A.: Prospect theory: an analysis of decision under risk. Econometrica **47**(2), 263–292 (1979)
14. von Neumann, J., Morgenstern, O.: Theory of Games and Economic Behavior. Princeton University Press, Princeton (1944)
15. Camerer, C.F.: Behavioural studies of strategic thinking in games. Trends Cogn. Sci. **7**, 225–231 (2003)

16. Barron, F.H., Barrett, B.E.: Decision quality using ranked attribute weights. Manag. Sci. **42**, 1515–1523 (1996)
17. Jia, J., Fischer, G.W., Dyer, J.S.: Attribute weighting methods and decision quality in the presence of response error: a simulation study. J. Behav. Decis. Mak. **11**, 85–105 (1998)
18. Mckelvey, R.D., Mclennan, A.M., Turocy, T.L.: Gambit: Software Tools for Game Theory. Version 0.2007.01.30 (2007). http://www.gambit-project.org
19. Jablonský, J.: MS Excel based system for multicriteria evaluation of alternatives. University of Economics Prague, Department of Econometrics (2009). http://nb.vse.cz/~jablon/
20. Johnson-Laird, P.N., Shafir, E.: The interaction between reasoning and decision making: an introduction. Cognition **49**(1–2), 1–9 (1993)
21. Milikkovic, D.: Rational choice and irrational individuals or simply irrational theory: a critical review of the hypothesis of perfect rationality. J. Socio-Econ. **34**(5), 621–634 (2005)

# What Computers Can Tell Us About Emotions – Classification of Affective Communication in Electronic Negotiations by Supervised Machine Learning

Michael Filzmoser$^{(\boxtimes)}$ [iD], Sabine T. Koeszegi [iD],
and Guenther Pfeffer [iD]

TU Wien, Institute of Management Science, Theresianumgasse 27,
1040 Vienna, Austria
{michael.filzmoser, sabine.koeszegi,
guenther.pfeffer}@tuwien.ac.at

**Abstract.** Affective communication and emotions are an important part of negotiations. Negotiation support and negotiation support systems, however, tend to neglect this aspect given extant measurement difficulties. This study explores the possibilities of state of the art supervised machine learning techniques to classify emotions expressed in negotiation communication during electronic negotiation experiments. The affective content of the exchanged messages was determined by human coders and classified according to the circumflex model of affect. The output of this laborious activity, that can only be accomplished after a negotiation, which makes it irrelevant for negotiation support, was input to this study. Promising performance of some preprocessing and machine learning techniques was achieved. Especially the category of activating negative emotions, which is highly important in negotiations as it might reduce the prospects of reaching an agreement, was correctly classified quite often.

**Keywords:** Affect · Electronic negotiations · Machine learning

## 1 Introduction

Despite a focus on analytic aspects in negotiation research, the significant role of affect in negotiation processes and outcomes has been acknowledged among negotiation scholars (for overviews see e.g. [1, 2]). Particularly a research team around Van Kleef and De Dreu has analyzed the impact of various emotions such as anger, happiness, worry, guilt, regret, disappointment, etc. on negotiation processes and outcomes (e.g. [2–5]). However, in electronic negotiation support research, affect and emotions have been understudied so far [6–8].

In computer-mediated communication, affectivity is the sensitivity to attitudes toward the communication partner or the subject matter in a communication and denotes the inclusion of affective components in a (text) message. Affect comprises emotions, which are directed towards specific situational stimuli, of shorter duration

© Springer International Publishing AG 2017
D. Bajwa et al. (Eds.): GDN 2016, LNBIP 274, pp. 113–123, 2017.
DOI: 10.1007/978-3-319-52624-9_9

and higher intensity and moods, which lack the quality of directedness but are more enduring and pervasive [2]. High affective complexity is associated with relational oriented obstacles such as mistrust and affective disruptions and therefore needs to be considered when deciding on the communication and negotiation strategy [9]. Especially in text-based negotiations affectivity can not only be indicative – i.e. consistent with, and thereby revealing, the affective state of a person – but also instrumental and therefore used strategically in the negotiation – e.g. expression of anger to elicit concessions [2].

Affect needs to be encoded or contextualized differently in computer-mediated communication compared to face-to-face communication with available non- and para-verbal cues. One possibility to contextualize emotions in text messages are emoticons (standing for emotion and icon) which are referred to relation icons, visual cues or pictographs and serve as surrogates for non-verbal communication to express emotion [10]. Additionally, communicators can use contextualization cues such as non-standard spelling, letter and punctuation mark repetition (e.g. '???') or lexical surrogates ('hmmm') and the like as linguistic form to express affect. All these cues contribute to the signaling of "contextual presuppositions" that allow for inferences about the meanings communicators intend to convey in a specific situation [11]. However, a substantial proportion of affective content is encoded implicitly in factual statements by communicators' lexical and syntactical choices. Not only what negotiators convey in their messages (content or substantial dimension) but also how they express themselves (affective dimension) substantially impacts the relationship and trust building between negotiators [12].

Te'ini therefore suggests a computerized support of communication strategies through e.g. templates of appropriate affectivity and feedback on current messages (e.g. language checks) [9]. Also Broekens et al. call for the development of negotiation support systems that also consider the affective dimensions [6]. The knowledge of the affective content of messages by negotiation support systems (NSS) or software agents would enable novel ways of supporting and automating negotiations. NSS could for example make the user aware of the affective content of own messages and messages of the opponent and thereby support the negotiator in a similar way to offer evaluation and generation [13]. Software agents could react not only to the offer behavior but also to affect explicated in messages of their human counterparts in semi-automated negotiations [14, 15].

This requires, first of all, the identification of affectivity in texts which is challenging because of the particularities of computer-mediated communications discussed above. In this paper we, therefore, focus on the identification and classification of affect in text-based negotiation messages by means of machine learning. The research question of this explorative study, therefore, is: "To what extent and in which quality are state of the art text preprocessing and supervised machine learning techniques able to classify affective communication in electronic negotiations?" To address this question we evaluate the performance of available techniques in supervised machine-learning, i.e. their ability to correctly assign electronic negotiation messages to the affective categories they were assigned to by human coders.

The remainder of this paper is structured as follows: Sect. 2 offers a brief theoretical background on affect classification, and Sect. 3 presents the data from electronic negotiation experiments applied in this data-driven approach. The negotiation case, the NSS applied and the coding and multi-dimensional scaling analysis to assign the messages to affective categories are also discussed in this section. Furthermore it introduces the preprocessing and supervised machine-learning techniques evaluated in this study as well as the experimental design. Section 4 presents and discusses the results of our study and derives suggestions for parameterization and algorithms for affect identification and classification in electronic negotiations. Section 5 concludes with a summary of the main findings and a discussion of future research.

## 2  Theoretical Background

An issue to resolve with regard to affect identification and classification is the potential complexity of emotion patterns. Even though [16] only differentiated between seven basic emotions (sadness, anger, happiness, contempt, fear, disgust, and surprise) hundreds of facets of emotions and emotion-related states have been identified in literature. We therefore suggest employing a dimensional model as suggested in [17]. In this two-dimensional perspective of affect, all emotions and emotion-related states can be represented by the two underlying bipolar affective dimensions of (i) valence (pleasure vs. displeasure) and (ii) degree of activation (high vs. low) [17–19] see Fig. 1 (adapted from [20: p. 141]).

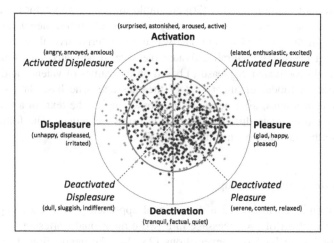

**Fig. 1.** Affectivity of messages in negotiations with (yellow) and without (blue) agreement (Color figure online)

In contrast to approaches based on discrete single emotions, which were often employed in previous work [21, 22], a dimensional approach provides a compact representation of the (implicit or explicit) "emotion quality" of each communication utterance in a two-dimensional Cartesian space and is preferable for the analysis of conversational settings [23, 24]. We therefore suggest identifying affect in negotiations by measuring the two dimensions of affective behavior, i.e. valence and arousal, manifest in communication behavior.

# 3 Data and Method

For our analyses, we used data from a previous negotiation experiment conducted with the NSS Negoisst [25], a web-based system that offers both decision and communication support. Participants in the negotiation experiment, students from negotiation courses of four European universities, represented either a Western European or an Eastern European company in, high conflict narrow zone of possible agreement, bilateral joint venture negotiations. The case contained seven issues with several continuous and discrete options and therefore was quite complex. The NSS recorded all exchanged offers and messages. A total of 57 negotiations between 114 negotiators were conducted from which 38 reached an agreement while 19 failed to reach an agreement. In all 57 negotiations a total of 730 messages were exchanged.

The messages from the electronic negotiations were first grouped according to their affective similarity by 26 unbiased business students. Each student received up to 250 of the 730 messages and working instructions that indicated that they had to build decks (no limit on the number of decks was provided) with similar messages. The similarity between messages – measured in the number of times these messages occurred in the same deck – was input to a multi-dimensional scaling analysis. These analyses were part of another study [20] and build the base data set for the analysis of possibilities of machine learning to identify affect in electronic negotiations in this study. The multi-dimensional scaling data was used to derive five affect categories for the negotiation messages (neutral, activated pleasure, deactivated pleasure, activated displeasure and deactivated displeasure) based on the values of valence and activation in the circumplex model of affect [17] of the message. The base data set of these categories and messages, after necessary prepossessing of the text, in a last step was used to train and evaluate different machine learning techniques. The following subsections describe the steps of the study in detail.

## 3.1  Multi-dimensional Scaling

As already mentioned in Sect. 2 a dimensional approach provides a compact representation of the affect of each message in a two-dimensional Cartesian space. This is preferable for the analysis of conversations [23, 24] like negotiations to approaches based on discrete single emotions. The evaluation of the similarity of the affective content of messages by human raters in a three step multi-dimensional scaling procedure builds the basis for the analyses of the subsequent sections.

In a first step the input data for multi-dimensional scaling is generated. For this purpose human raters evaluated the affective similarity of the negotiation messages exchanged. 26 business students participated in this rating activity, they received no background information about the underlying study but detailed instructions to rate each up to 250 of the 730 messages. The task of the raters was to sort similar messages into the same deck. For this task the raters received no additional training or instructions, like coding schemes, number of decks, etc.

This data built the basis for multi-dimensional scaling based on the proximity of two messages, which was measured by the number of raters who assigned them to the same deck. The proximity matrix was processed by the multi-dimensional scaling software PERMAP 11.8a using nonparametric multi-dimensional scaling with Euclidean distances as distance measures, the preferable approach for proximity measures based on subjective judgments [26].

Goodness of fit (Stress-1) and the interpretation of possible dimensions indicated that a two-dimensional Cartesian space best fitted the data. Rotation of the axes lead to the two dimensions of valence and activation, which bring the results of the multi-dimensional scaling in accord with the circumplex model of affect [17]. A detailed description of the multi-dimensional scaling procedure can be found in [27].

For the categorization task of supervised machine-learning the data has to be distinguished and labeled into discrete classes which were determined according to the four sectors of the circumplex model of affect plus a neutral class which contains messages in the center and therefore of low affectivity. This resulted in approximately equal amounts of messages in all five classes as represented in Fig. 2.

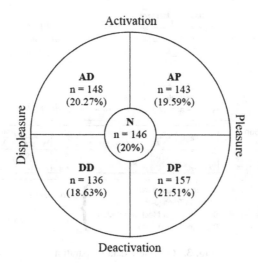

**Fig. 2.** Classes of affective communication and number of observations.

The adequacy of the existing variety of techniques for data preprocessing and supervised machine-learning for classification of affective content of electronic negotiation messages at present is unclear. There are some promising applications of machine-learning for sentiment analysis, i.e. the evaluation of whether there is a general positive or negative feeling towards an issue from blogs, newspapers, forums or stock reports [28]. However, there are significant differences between sentiment analysis and analysis of the affectivity of electronic negotiation messages, which hinder the direct application of these methods to the field of electronic negotiations. On the one hand the available amount of data from forums, blogs, etc. for machine-learning is considerably larger, on the other hand for sentiment analysis the classification need not be as detailed, a general tendency is sufficient while more concrete classifications of dimension of affectivity are considered in this study. This calls for a systematic comparison of the available options both for data preprocessing and machine-learning. We subsequently briefly describe these techniques, which are all implemented in WEKA, an open source software that implements various machine-learning techniques for (big-data) data-mining purposes, which was applied for the analyses of this study.

## 3.2   Data Preprocessing Techniques

Identification of emotions in electronic negotiation messages is basically a text mining task. Text mining is a variant of data mining. However, the data used in data mining is usually more structured than communication transcripts. To make the established algorithms from data mining available to text mining the complexity and variety of the data has to be reduced. A variety of techniques exist for this purpose. Figure 3 gives an overview of the data preparation.

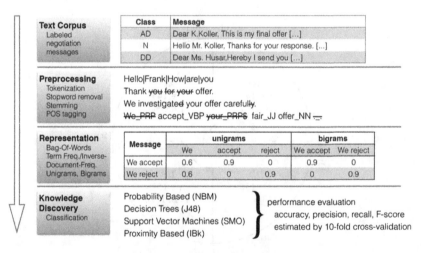

**Fig. 3.**  Overview data preparation

Stemming is the reduction of different times and forms of words to their roots, like e.g. 'is', 'was' and 'am' to 'be'. This complexity reduction technique thereby improves the performance of word mining algorithms as it increases similarities. Stopword removal eliminates the most often used words ('the', 'a', 'and', etc.) from the data set These stopwords are often equally present in all classes so that they do not add to the discriminative power of an algorithm but are rather noise, which can be filtered out with stopword removal. Stopwords can be taken from general lists (e.g. Swish-E) or be the most frequent (10, 50, 100, etc.) words in the data set. N-grams are word combinations, e.g. of two words like 'hello dear', 'hello sir', or 'dear sir' in the case of bi-grams, rather than just single words, like 'hello', 'dear' or 'sir' and are often more informative for classification algorithms than the single words. Part of speech (POS) tagging categorizes the original text into grammatical classes e.g. verbs, nouns and adjectives and replaces the words by these classes to reduce variance and facilitate pattern recognition.

## 3.3 Machine-Learning Techniques

After application of the data preprocessing techniques, to reduce variance and facilitate classification, machine-learning algorithms perform the actual classification task. Naive Bayes is a probabilistic classifier based on Bayes' theorem. The hypothesis that an object belongs to a certain class is updated by learning and the actual assignment of an object to a class bases on the probability that the different hypotheses are true. Decision tree algorithms develop an internal hierarchy of nodes and arcs, where the former represent decision points and the latter stand for the classes. The trained model is the result of creating a tree with maximum discriminatory power of each decision node. The support vector machine approach fits during the training phase mathematical functions to the multi-dimensional feature space which are then used for classification. Proximity approaches like (k-) nearest neighbor are 'lazy' learning approaches. The (k) most similar objects from the – already classified – training data set to the focal object of the test set are identified and the test object is assigned to the most frequent class elicited this way. Machine-learning algorithms are trained on a training data set and then tested on a test data set. In this study we separate the total data set of 730 messages in ten subsets from which in ten runs each subset is used as test data set with the remaining nine data sets as training data set.

## 3.4 Experimental Design and Measurement

The data preprocessing techniques are combined to a total of 16 experimental settings (#01 to #16), illustrated in Table 1. The resulting data sets are then input to the four machine-learning techniques discussed above: Naïve Bayes (NBM), decision tree (J48), support vector machines (SMO) and nearest neighbor (lBk).

For the comparison of the performance of the machine-learning algorithms we apply the four measures suggested in [29]. The accuracy of an algorithm (1) is the percentage of correctly identified messages.

$$accuracy = \frac{correct}{total} \qquad (1)$$

Accuracy, however, does provide little information about the characteristics of the classification errors. An algorithm can correctly (true) assign a message that belongs to the focal class to this class (true positive) or not to other classes (true negative), as well as it can incorrectly (false) assign a message that belongs to another class to the focal one (false positive) or that belongs to the focal class to other classes (false negative). Based on these correct and incorrect classifications and error types the precision (2) and recall (3) rations of the algorithm can be determined.

**Table 1.** Experiment settings – data preprocessing techniques.

| Setting | Dataset | n-grams | Stopwords | Stemmer |
|---------|---------|---------|-----------|---------|
| #01 | original | unigram | none | none |
| #02 | original | unigram | none | Porter |
| #03 | original | unigram | Swish-E | none |
| #04 | original | unigram | Swish-E | Porter |
| #05 | original | unigram | Top 50 | none |
| #06 | original | unigram | Top 50 | Porter |
| #07 | original | uni- & bigram | none | none |
| #08 | original | uni- & bigram | none | Porter |
| #09 | original | uni- & bigram | Swish-E | none |
| #10 | original | uni- & bigram | Swish-E | Porter |
| #11 | original | uni- & bigram | Top 50 | none |
| #12 | original | uni- & bigram | Top 50 | Porter |
| #13 | POS adjusted | unigram | none | none |
| #14 | POS adjusted | unigram | Top 50 | none |
| #15 | POS adjusted | uni- & bigram | none | none |
| #16 | POS adjusted | uni- & bigram | Top 50 | none |

$$precision = \frac{true\,positive}{true\,positive + false\,positive} \tag{2}$$

$$recall = \frac{true\,positive}{true\,positive + false\,negative} \tag{3}$$

Precision and recall are interdependent, therefore, the so-called f-score (4) is used as a harmonic mean of precision and recall, which is typically used as overall performance measure for machine-learning algorithms.

$$f - score = \frac{2 \times precision \times recall}{precision + recall} \tag{4}$$

As a conservative baseline for the evaluation of the accuracy of the machine-learning algorithms we define the percentage of the largest class (DP deactivation-pleasure) with 21.51% of the messages. This accuracy would be achieved by a plain algorithm that assigning all messages to the largest class, a random assignment to one of the five classes would lead to a slightly more generous baseline of 20% only.

## 4    Results

As can be seen from Fig. 4 all machine-learning techniques are significantly above the comparison baseline of 21.51% accuracy, which would be achieved by assigning all messages to the class DP which has the highest share. Moreover, on the one hand significant performance differences between machine-learning techniques exist and on

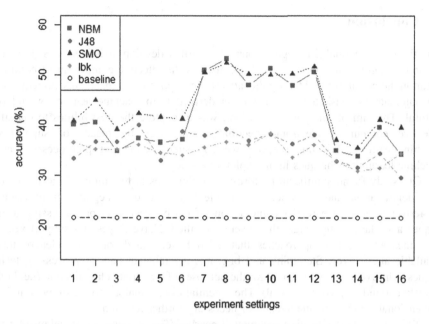

**Fig. 4.** Classification accuracy overview.

the other hand significant interaction effects exist between machine-learning and pre-processing techniques that have a major impact on performance.

Support vector machines and Naïve Bayes perform approximately equal and better than decision tree or nearest neighbor approaches. Furthermore they are especially good when combined with bi-grams, stemming and no stopword removal or POS adjustment.

Besides this overall performance to correctly classify all five classes the detailed performance per class is also of interest. Especially the activating negative emotions are critical for negotiation success as they might lead to negotiation break-offs. Table 2 presents the detailed classification results for all five categories of the two best performing machine-learning techniques (Naïve Bayes and support vector machines) combined with the best performing data preprocessing approaches (#08: no POS adjustment, usage of uni- and bi-grams, no stopword exclusion and usage of a stemmer).

**Table 2.** Detailed results for algorithms.

| Class | #08.NBM | | | #08.SMO | | |
|---|---|---|---|---|---|---|
| | Precision | Recall | f-score | Precision | Recall | f-score |
| N | 63.6% | 43.2% | 51.4% | 50.7% | 47.3% | 48.9% |
| AP | 53.9% | 48.3% | 50.9% | 55.6% | 51.7% | 53.6% |
| AD | 62.4% | 56.1% | 59.1% | 62.2% | 53.4% | 57.5% |
| DD | 44.4% | 58.1% | 50.3% | 44.5% | 53.7% | 48.7% |
| DP | 49.5% | 60.5% | 54.4% | 51.8% | 56.1% | 53.8% |
| Weighted avg. | 54.8% | 53.3% | 53.3% | 53.1% | 52.5% | 52.6% |

## 5  Conclusion

Emotions are crucial in negotiations. To further-develop existing NSS towards proactive negotiation support that also considers the affective component of communication the possibility to identify this affective component is a mandatory prerequisite. To automate this task the knowledge set derived from machine learning would be helpful. The aim of this paper, therefore, was to explore the possibilities to classify affective content of electronic negotiation messages by means of supervised machine-learning. For this purpose we compared extant text preprocessing and machine-learning techniques in an explorative study.

Our study found significant performance differences of text mining algorithms for the identification and classification of affect in electronic negotiation messages. Moreover performance relevant interaction effects between data preprocessing techniques and classification algorithms were identified. Naïve Bayes and support vector machines are the two approaches that seem better suited for this endeavor than available alternatives. Stemming and bi-grams are relevant data preprocessing techniques, while others are not suited for the purposes of affectivity classification (i.e. POS adjustment and stopword removal). The potential discriminatory power of even higher dimensional n-grams is one avenue of necessary further research.

The achieved classification accuracy of nearly 55% is a promising initial result, and the accuracy of over 62% for the important activating negative category even more so. However, the performance is still not satisfying for the actual implementation in NSS. Additional research is necessary to achieve this ultimate goal. Especially more coded data, which is laborious work, and more textual indicators for affect in negotiation messages are necessary to establish a convenient training data set. The 730 messages from 57 negotiations are a relatively small data set compared to the 'big data' problems for which machine-learning is usually applied. This data should also be more diverse, featuring different levels of conflict different types of negotiations etc. to avoid over-fitting for a specific negotiation problem.

## References

1. Kumar, R.: The role of affect in negotiations: an integrative overview. J. Appl. Behav. Sci. **33**(1), 84–100 (1997)
2. Barry, B., Fulmer, I.S., van Kleef, G.A.: I laughed, I cried, I settled: the role of emotion in negotiation. In: Gelfand, M.J., Brett, J.M. (eds.) The Handbook of Negotiation and Culture, pp. 71–94. Stanford University Press, Palo Alto (2004)
3. van Kleef, G.A., de Dreu, C.K.W., Manstead, A.S.R.: The interpersonal effects of anger and happiness in negotiations. J. Pers. Soc. Psychol. **86**(1), 57–76 (2004)
4. van Kleef, G.A., de Dreu, C.K.W., Manstead, A.S.R.: The interpersonal effects of emotions in negotiations: a motivated information processing approach. J. Pers. Soc. Psychol. **87**(4), 510–528 (2004)
5. van Kleef, G.A., van Lange, P.A.M.: What other's disappointment may do to selfish people: emotion and social value orientation in a negotiation context. Pers. Soc. Psychol. Bull. **34**(8), 1084–1095 (2008)

6. Broekens, J., Jonker, C.M., Meyer, J.J.: Affective negotiation support systems. J. Ambient Intell. Smart Environ. **2**(2), 121–144 (2010)
7. Johnson, N.A., Cooper, R.B., Chin, W.W.: Anger and flaming in computer-mediated negotiation among strangers. Decis. Support Sys. **46**(3), 660–672 (2009)
8. Martinovski, B.: Emotion in negotiation. In: Kilgour, D.M., Eden, C. (eds.) Handbook of Group Decision and Negotiation, vol. 4, pp. 65–86. Springer, Netherland (2010)
9. Te'ini, D.: A cognitive-affective model of organizational communication for designing IT. Manag. Inf. Sys. Q. **25**(2), 251–312 (2001)
10. Walther, J.B., D'Addario, K.P.: The impacts of emoticons on message interpretation in computer-mediated communication. Soc. Sci. Comput. Rev. **19**, 324–347 (2001)
11. Darics, E.: Non-verbal signalling in digital discourse: the case of letter repetition. Discourse Context Media **2**, 141–148 (2013)
12. Griessmair, M., Koeszegi, S.T.: Exploring the cognitive-emotional fugue in electronic negotiations. Group Decis. Negot. **18**(3), 213–234 (2009)
13. Vetschera, R., Filzmoser, M., Mitterhofer, R.: An analytical approach to offer generation in concession-based negotiation processes. Group Decis. Negot. **23**(1), 71–99 (2014)
14. Filzmoser, M.: Simulation of Automated Negotiation. Springer, Vienna (2010)
15. Filzmoser, M.: Automated vs human negotiation. Int. J. Artif. Intell. **4**(10), 64–77 (2010)
16. Ekman, P., Cordaro, D.: What is meant by calling emotions basic. Emot. Rev. **3**(4), 364–370 (2011)
17. Barrett, L.F.: Feelings or words? Understanding the content in self-report ratings of experienced emotion. J. Pers. Soc. Psychol. **87**(2), 266–281 (2004)
18. Russell, J.A.: A circumplex model of affect. J. Pers. Soc. Psychol. **39**(6), 1161–1178 (1980)
19. Watson, D., Tellegen, A.: Toward a consensual structure of mood. Psychol. Bull. **98**(2), 219–235 (1985)
20. Hippmann, P.: Multi-level dynamics of affective behaviors in text-based online negotiations: impacts on negotiation success and impacts of decision support. Doctoral thesis, University of Vienna (2014)
21. Lazarus, R.S., Smith, C.A.: Knowledge and appraisal in the cognition—emotion relationship. Cogn. Emot. **2**(4), 281–300 (1988)
22. Ortony, A., Clore, G.L., Foss, M.A.: The referential structure of the affective lexicon. Cogn. Sci. **11**(3), 341–364 (1987)
23. Burgoon, J.K., Hale, J.L.: The fundamental topoi of relational communication. Commun. Monogr. **51**, 193–214 (1984)
24. Frijda, N.H.: Emotions, individual differences and time course: reflections. Cogn. Emot. **23**(7), 1444–1461 (2009)
25. Schoop, M., Jertila, A., List, T.: Negoisst: a negotiation support system for electronic business-to-business negotiations in e-commerce. Data Knowl. Eng. **47**(3), 371–401 (2003)
26. Heady, R.B., Lucas, J.L.: Permap: an interactive program for making perceptual maps. Behav. Res. Methods Instrum. Comput. **29**(3), 450–455 (1997)
27. Filzmoser, M., Hippmann, P., Vetschera, R.: Analyzing the multiple dimensions of negotiation processes. Group Decis. Negot. **25**(6), 1169–1188 (2016). doi:10.1007/s10726-016
28. Devitt, A., Ahmad, K.: Is there a language of sentiment? An analysis of lexical resources for sentiment analysis. Lang. Resour. Eval. **47**(2), 475–511 (2013)
29. Weiss, S.M., Indurkhya, N., Zhang, N.: Fundamentals of Predictive Text Mining. Springer, London (2010)

# Applications of Group Decision and Negotiation

# Facebook and the Elderly: The Benefits of Social Media Adoption for Aged Care Facility Residents

Saara Matilainen[1], David G. Schwartz[2,3] ⓘ,
and John Zeleznikow[4,5(✉)] ⓘ

[1] College of Business, Victoria University, Melbourne, Australia
saara.matilainen@live.vu.edu.au
[2] Graduate School of Business Administration, Bar-Ilan University,
Ramat Gan, Israel
david.schwartz@biu.ac.il
[3] Visiting Scholar, College of Business, Victoria University,
Melbourne, VIC, Australia
[4] Centre for Cultural Diversity and Wellbeing, Victoria University,
Melbourne, VIC, Australia
john.zeleznikow@vu.edu.au
[5] Bar Ilan University, Ramat Gan, Israel

**Abstract.** We explore the emotional effects of implementing Facebook in an aged care facility and evaluate whether computers and Facebook are of any benefit in regard to an elderly person's feeling of social connectedness. This preliminary qualitative study took place in a Melbourne-based elected Aged Care Facility. Facebook was accessed through desktop computers provided by the Facility.

During a four month pilot project, six residents were supervised to learn how to competently Facebook. Findings indicate that older people are able to connect and learn the use of new technologies with which they may be unfamiliar. While high levels of user enjoyment were found, measures of social connectedness resulting from the use of Facebook use were inconclusive. The research concludes that when supported with appropriate teaching and technology, the use of computers to access Facebook is a practical approach in supporting the connectedness needs of the residents.

**Keywords:** Social media · Elderly · Connectedness · Isolation

## 1 Introduction

The Western World's aging population is growing rapidly, thus leading to considerable stresses on the health care systems [1–3]. Globally the trend of decreasing mortality rates and the decrease in fertility rates since the 1950's has led to an aging population. This is referred to as a demographic transition [4: p. 5].

In Australia, it is predicted that the next 50 years will see drastic changes in the age structure of the population. These changes are reflected in the growth of the median age

© Springer International Publishing AG 2017
D. Bajwa et al. (Eds.): GDN 2016, LNBIP 274, pp. 127–139, 2017.
DOI: 10.1007/978-3-319-52624-9_10

from 37.7 years in 2012, to between 38.6 and 40.5 years of age in 2031. The age group of 65 years and over is projected to rise from 3.2 million in 2012, to between 5.7 and 5.8 million in 2031 [5].

## 1.1 Aging and Socialisation

Getting older can be a difficult time for many individuals with the onset of physical and mental ailments and the prospect of relocating to an aged care facility when one cannot be adequately cared for at home. This adjustment can be a daunting task and many individuals worry about what will happen at this stage of their lives. Previous research points out that once a person moves to a care type facility, there is a reduction in their quality of life along with the reduction in their independence and difficulties with socialisation [6, 7]. Subsequently, social isolation and loneliness are common. Vast social problems are found in nursing home settings and seem to contribute to deteriorating physical and mental health of residents [8–11].

Loneliness is experienced by 45% of Australians aged 65 years and over [12]. The prevailing social environment is an important factor which correlates social behaviour and physical health [13]. Ongoing socialisation has been observed to maintain patients' mental state, and often prevents the onset of such complaints and ailments as loneliness, and depression. The lack of social interaction leads to increased rates of morbidity and mortality [8–10, 13–15].

Research on social isolation has shown that ongoing socialisation is a vital part of slowing the aging process and has many benefits for the individual. Characteristics of intervention methods that appear to be effective at targeting social isolation include offering social activities to older people, offering support within a group format, and offering activities where people are required to actively participate [3]. Many studies demonstrate that few intervention methods for social isolations have been successful and many have failed to have long lasting effects [3, 13].

## 1.2 Aging, Information Technology and Social Media

Although it is thought to be common knowledge that older people, generally above the age of 65 have a lack of understanding of modern technology including the internet, it should not be assumed that older people do not want to learn the use of new technology [6]. Data found on the internet use of persons aged 65 years or older demonstrates that in 2012–2013, 46% of Australian older persons were internet users [16].

Many preceding studies demonstrate the prospect of Information and Communications Technology (ICT) adoption in nursing home settings. These studies demonstrate the benefits associated with ICT use and improvements in mental health and care provided. Some of the pitfalls in these research studies were that at the time of the research the technology was not advanced enough to facilitate long distance video calls or to perform multiple tasks. This led to frustrated users because of reoccurring complications whilst using the technology. Hence there has been a delay in implementing this technology in aged care settings. [9, 10]. Today we are able to avoid most of these hurdles as internet, smart technologies, and applications have advanced to accommodate user requirements.

An Australian national survey of 800 senior participants aged 55 and over reported that access to the internet made them feel more connected and helped them overcome feelings of social isolation and loneliness. The seniors used the internet to connect with their children and grandchildren as well as other family members. The findings demonstrated the ability to adapt and adopt the use of the technology, as well as reporting that one third of seniors used video calling applications such as FaceTime or Skype to stay in touch with family and friends and that more than half used Facebook and email [12].

According to Technology Review, in the future technology will play a major role in reducing social isolation and loneliness in aged care services accommodation. This trend will be significantly higher in families that are separated by long distances or who live in remote areas [12]. Egan's research confirms future possibilities in fostering communication in nursing home settings through ICT. Our research has taken the next step and trialled Facebook in an aged care facility, with a positive outcome, reinforcing the ideal of this technology assisting the elderly in the pursuit of finding a solution for social isolation. It also highlights a new user group, and the customisation required to meet the needs of this group.

But can those we define as elderly, benefit from the use of social media and information technology? Bell et al. [17] claim that social media has been widely adopted by younger adults, but older adults have been less likely to use such applications.

They conducted a survey of 142 older adults ($Mage$ = 72 years, $SD$ = 11; range: 52–92) living in the metropolitan Atlanta area to understand the characteristics of older adults who do and do not use Facebook. They examined the relationship between Facebook use and loneliness, social satisfaction, and confidence with technology. Fifty-nine participants (42%) identified themselves as current Facebook users; 83 participants (58%) were not Facebook users. Non-Facebook users were significantly older ($Mage$ = 75.3 years) than Facebook users ($Mage$ = 66.5 years). Counter to expectations, there was not a significant difference in loneliness between Facebook users and non-users for this sample. However, Facebook users did score higher on assessments of social satisfaction and confidence with technology than did non-users.

Their preliminary results suggest that many older adults do use Facebook and they primarily use it to stay connected with family. They claim that social media may begin to play a more active role in keeping an ageing population socially connected. Therefore, understanding the factors that influence social media use in older adults is becoming more critical.

Hope et al. [18] present results from an interview study involving 22 older adults (age 71–92) to understand communication preferences and values related to social media. They studied older adults who are interested in improving communication with their family and friends. But they noted that these seniors rarely use social media to connect with members of their social networks. As a whole, older adults value deeper, well thought out, carefully crafted social communications that are achieved through telephone calls, e-mails, and written letters. Issues of privacy, information credibility, and content relevance are key reasons for not using social media.

They claim that participation on social media sites requires a time commitment and there are expectations of reciprocity. Encouraging seniors to connect with weak ties may deter use. They argue that while older adults perform many social functions that could be supported by online technologies, few seniors use such systems.

Hutto and Bell [19] claim that whilst the percentage of older adults using social media has dramatically increased in recent years, comparatively little research has been done to understand this unique community of users. They explore several characteristics of active Facebook users among older adult and build on previous research to investigate the differential impact of traditional versus social media-mediated communication activities among older adults, and assess its relationship with social satisfaction.

They then examine the specific relationship between older adults' Facebook communication habits and their attitudes regarding social satisfaction, loneliness and social isolation. Controlling for factors such as age, gender, ethnicity, socioeconomic status (education and income), and marital status, they find that directed communications (as opposed to broadcast communications and passive consumption of content) is significantly correlated with feelings of social satisfaction among this distinct population.

They also found a significant relationship between age and network size - older seniors had distinctly smaller social networks than younger seniors. A strong correlation between smaller networks and increasing age has implications for the study of social networks in general. For example, as the locus of social networks begins to shift away from colleagues and other workplace acquaintances more towards family and close friends, one might postulate that older adults may begin to develop stronger ties to members of their shrinking network. If the strength of ties between older adults and their network members is significantly stronger than the ties between younger adults in otherwise comparable networks, then this has meaning for a broad range of research interests that are based on social network simulation models or general studies of the diffusion of innovations. When Hutto and Bell examined the strength of the relationship between age and the size of social networks among older adults, what they were not able to answer here is whether older adults with fewer connections actually have stronger connections among their network ties.

Among older adult social media users, they found no differential effects of social media-based communications versus traditional communication channels with regards to social satisfaction. Seniors appear to use social media communication to supplement traditional forms of communication without impacting their social satisfaction. However, when contrasted against other Social Network Sites specific communication activities, older adults with more directed communications per year had significantly higher satisfaction with their own role and activities within their social networks. Direct interactions are comparatively more effortful than broadcast communications or passive consumption, but these interactions are a simple and convenient way to remain engaged in at least some part of another's daily life. Even the lightest of lightweight interactions can signal that the person feels that a relationship is meaningful – an important part of building and maintaining strong social connectedness for seniors.

Previous research has investigated the use of information technology and social media by the elderly, and how social media can help enhance connectedness. But very little work has been performed on the combination of these two topics. We address this issue in this paper.

Given the age of our participants, it would not have been appropriate to conduct quantitative research requiring the completion of surveys. Instead, we decided upon conducting a small qualitative study.

Observing the issues related to a lack of socialisation became the main identifier for the problem. This research studies the impact of introducing Facebook to support social interaction. It focuses on the five features of; Participation, Openness, Conversation, Community, and Connectedness [20]. This led to several factors motivating the research including, the need to understand whether technology can assist older people, why older people do not use smart technology, and the desire to gain an in-depth understanding of how the target group uses social media.

This preliminary research addresses the following goals:

1. *To discuss complications and difficulties that arise in relation to the use of computers and Facebook by our target population, and*
2. *To measure how Facebook helped connect the elderly with family and friends and in turn assisted with their perceived feelings of loneliness and social isolation.*

## 2  Methods

The study was carried out at a Melbourne Jewish Aged Care Facility[1] that caters for individuals with both high and low care needs. The majority of the elderly people who reside at this aged care facility speak Russian as their native tongue and have a limited use of the English language. Ideally, we would have chosen native English speaking residents for our trial. But very few native English speakers live at the St Kilda Road residence of the nursing home – the site of our experiment.

Participants for this study were selected based on their willingness to participate, having no or limited mental and visual impairment, the physical ability to use a computer, reasonable proficiency in English and having family with whom they could communicate via social media.

Initially, the facility manager selected 11 capable residents for the study. It was then up to the research team to recruit the participants. The reason that only a small portion of residents were asked to participate was due to time and resource limitations. Initially all 11 people recommended by the facility manager were asked to participate in the research; 6 of whom agreed and were interviewed for the first set of interviews and only 3 of whom remained until the completion of the research project. Reasons for-non completion include one participant passing away, unavailability to be interviewed for the second interview, and finally one participant felt they were not able to participate in the trial of the technology due to fear of lack of knowledge and limited English language. Out of the 11 prospective participants, reasons given for declining involvement in the project were due to some residents being immigrants to Australia from a non-English speaking background and hence not feeling proficient in English.

---

[1] See http://www.jewishcare.org.au/residential-aged-care/ last accessed November 19, 2016.

Whilst all the relevant documents were translated into Russian[2], many native Russian speakers were reluctant to be involved. Hence there were only a small number of participants in the project. This also raises the issue of culture (as opposed to mere translation) as an important factor in the use of Social Media. The willing participants ranged in age from 69 to 88, and all of them resided at the aged care facility alone, having no close relative or spouse live with them at the facility. Two sets of open ended interviews were used to gather the data, one before the implementation of Facebook and the second after the implementation of Facebook. Facebook was chosen as the preferred form of social media due to its popularity.

To begin, the research participants were interviewed individually in an hour long consultation. The interviews consisted of 20 open ended interview questions. The questions focused on determining how the residents felt about the use of computers, as well as use, familiarity, and attitudes toward one form of social media. The interviews were aimed at building a framework to understand the capability and prospect of implementing technology and social media into the aged care facility. The participants were also asked if they felt lonely and isolated and whether they believed their social needs were met by either the hosted activities at the facility or via family member visitations.

The second set of interview questions, conducted six months later, after participants received support on computer usage, the internet and social media, focused on gaining information regarding whether the participants perceived a changed level of socialisation after the implementation of social media, whether they enjoyed the experience and whether they found it useful. All of the participants opened up about their personal lives and candidly explained their experiences to the researcher. It should be noted that the questions were not based on a psychological evaluation of the resident, nor were they aimed to scientifically identify a resident's mental health.

## 2.1  Intervention

After the first sets of interviews were conducted and the level of socialisation and loneliness of the interviewee as well as his/her know-how of technology established, each participant was individually introduced to computers and given lessons developing their computer literacy to a sufficient level to enable them to gainfully use Facebook. The level of training depended on each person's individual needs. The computer lessons were provided by the researcher alone.

A total of 3 lessons each were given to each participant. The lessons varied in length from 1 to 2 h. The participants did not want to participate in a group meeting, fearing their lack of skills and knowledge would be critiqued by others at the residence. Some residents were familiar with the use of computers where-as others had never used a computer before. That is why each lesson was tailored to suit the needs of the resident. The lessons ranged from teaching residents how to open a computer, teaching them the use of the mouse and keyboard, teaching how to click and double click, to

---

[2] By official translators who were paid by Jewish Care Victoria.

providing support for participants with a more advanced knowledge of computers. This latter group knew how to type using a keyboard, how to open and close a computer and how to gain access to the internet. The lessons were provided for each resident to aid them first in creating an email address and second to assist them with the creation of a Facebook account.

Once the lessons were completed the researcher left the facility for a time period of four months. The participants were expected to use the computers and Facebook during this time, period without the assistance of the researcher. No other assistance was available for the participants in the study due to lack of resources. The researcher returned to the facility to ask the final set of interview questions. The data from the first and second set of interviews were compared and analysed using an explanation building method to identify what difference the technology had made and how it affected the resident's lives. These personal encounters were recorded and the results generated from the participants responses.

## 3  Results

### 3.1  Results of the First Set of Interviews

The first set of interviews revealed that all of the participants had heard of social media and 5 out of 6 were able to describe what it was. Only 2 participants out of the 6 had used social media previously and 5 out of 6 participants had used a computer or a smart phone or a tablet previously. All participants were excited, and or positive about the potential to learn to use computers and one form of social media. The first sets of interviews were indicative that all of the participants missed their families and felt lonely or isolated at times. About 80–90% of the residents at this aged care facility spoke a language other than English as their native tongue.

The participants who completed this particular study spoke English as their native tongue, thus the participant group expressed that they felt isolated from the rest of the residents, who were not native English speakers. Many of the activities hosted at the facility were tailored for non-English speakers. All of the participants felt that they could not talk to or engage with other residents at the aged care facility due to language and cultural barriers. Some other communication barriers included cognitive difficulties. The participants stated that at times they felt isolated due to the lack of English speaking residents at the facility, and because they were not able to engage or communicate with the majority of residents. Even though the Aged Care Facility provides many outings throughout the month and has several daily in-house activities, the participants felt they either could not join in on these activities because they felt they did not belong and they felt that the games provided, or the people playing the games, were below their own intellectual ability.

All of the participants spoke on the phone to their families several times a week and more than half were able to see family members at least once a week. Even though the level of communication between the residents and their relatives was moderate to high, four participants confessed to feeling very lonely and isolated, wishing they had someone to talk to at the residence or in person. All of the participants stated that they avoided contact with others on purpose, by not joining group meal times for example breakfast, lunch, or dinner and by staying in their rooms.

## 3.2    Computer Use

It proved challenging to create an email account for two of the participants, Using the mouse and keyboard was complicated due to the participants' shaky hands. The most difficult task participants found was choosing and typing the passwords as many mistakes were made whilst typing, and many requirements were needed for the creation of passwords, for example the combination of lower case and upper case letters, and numbers.

Creating a Facebook account was less challenging as most of the participants had learned basic computer use by this stage and had created an email address. One participant was not able to create a Facebook page because they did not master computer usage, nor could they create an email account. Creating the Facebook account required less information to be completed and did not require a combination of capital and smaller case letters. Each participant took about 10–15 min to create a Facebook account in comparison to establishing an email account which varied from 30–45 min. Each user name and password was recorded by the on duty facility Manager, as well as given to each participant on a piece of paper. Notices and information regarding how to access email and Facebook were laminated and hung on the walls of the computer room.

Out of the 6 participants 2 participants had existing Facebook accounts and 2 participants successfully created Facebook accounts for themselves. The oldest participant was not able to create a Facebook page because he needed more time to become familiar with computer use. This participant had never used a computer before and therefore needed more lessons to be able to create both an email account and a Facebook account. This participant was very keen to learn and showed a great interested in creating a Facebook profile and an email account. This particular participant felt he did not want to fall behind in the world of technology. The youngest participant was 69 years old, very mentally alert and learnt computer literacy and the use of email and Facebook the quickest.

Clearly not all elderly (defined as over 65) have the same skills. Current research classifies the elderly into three groups: 65–74, 75–84 and 85 and above [21].

## 3.3    Results of the Second Set of Interviews

The second set of interviews aimed to reveal the impact that the trial had on the participants. The interviews revealed that during the months the researcher was away, the organisation underwent a change and the computer laboratory was altered to accommodate for a wellness centre. According to the participants, the renovations began a couple of weeks after the researcher had left and they were not able to access the computers which had been placed there. This change largely impacted the influence computers and Facebook potentially could have had on socialisation, therefore altering the initial questions for the second set of interviews. This change also demonstrates how important management support is to enable residents to become familiar with computers and social media.

The final set of interviews did reveal some great insights into the wants, needs and attitudes of the elderly in regard to Facebook and computer usage. The 3 remaining participants uncovered that learning the use of computers was something they thoroughly enjoyed and being able to connect with old friends and family was a delight and something they thought they would never experience via a technological device. Comments included:

- *"I really enjoyed the experience and I think this sort of thing is very useful and interesting";*
- *"Computers and in particular Facebook helped me feel more connected to the outside world this kind of technology should be implemented on a permanent basis";*
- *"I am very eager to lean; if someone would help me I would definitely be interested in a program like this in the future";*
- *"I think the major issue was that no one took over the program once the facilitator left";*
- *"I am very upset that the changes took place, I really enjoyed the computer lessons with the facilitator, and I felt that they were very useful and fun.";*
- *"I would definitely participate in something like this again. I enjoyed using my Facebook page with the facilitator, but I was not able to use it on my own, I needed more practise".*

The two participants who already had social media accounts told the researcher they used the platform on a daily basis to connect and talk to their immediate family members. They stated that social media helped them stay in touch and that it was easy to use from the comfort of their own rooms. All of the participants interviewed stated that they would recommend computers and Facebook to other residents and hoped that this would become a permanent feature of the aged care facility. One participant made suggestions for the facility to place the computers in a common area so that residents would be able to have access to them on a weekly basis.

## 4 Discussion

More research is needed to understand the benefits and social implications of how the elderly use social media. This preliminary research illustrates the possibilities of Facebook enabling communication given the importance of retaining remaining social ties and the growing issue of social isolation in the elderly in an aged care facility. Whilst this study is merely a pilot study for future research, it aimed to discern whether a project of this kind is feasible. Our social welfare partners (Jewish Care Victoria), claim that the most isolated elderly are those living as individuals at home. And governments are keen to keep the elderly at home, rather than institutions, for as long as possible. So it would have been ideal, to survey this cohort. The limited resources available for the project, limited our ability to take such action yet pilot results are encouraging and appear to justify expanding the research in that direction.

A weakness of this research is the small number of participants which the researcher was able to gather for the purposes of this paper. However, given that this report is about a pilot study, where our goal was to find whether future research on the project would be viable, we were not concerned that the results might not be generalizable.

According to the findings of this paper, Facebook can be adopted and used by the elderly. This is conditional based on their level of cognitive comprehension, and mobility, and given they need to be provided with lessons and personal assistance. The small sample that participated in the research demonstrated positive attitudes toward the adoption of new technology. Facebook enabled two participants to connect with their family and friends on the social networking site and concluded that the two participants had a pleasant and enjoyable experience using the technology.

The research also discovered that two of the participants at the residential care facility had existing Facebook accounts, further adding to the possibility of the elderly using these applications. As time passes, a greater percentage of the elderly will have Facebook accounts.

It is important to mention that once the participants were denied access to the computers and no assistance was available for the participants to continue to access the technology, the participants became irritated because they wanted to continue learning and using the technology. Therefore it can be concluded that computers did provide a satisfying and practical experience to this group of people. Further, we learned from this project, that for the successful implementation of a social media it is imperative to provide appropriate and extensive training for those elderly who are commencing the use of computers and new technology.

It is vital that the different levels of comprehension and aptitude of residents are understood by the organiser of the training. Each participant in this study revealed that they comprehend the technology at varying rates and learn the technology at a different pace. The research team admits that more training should have been provided for each participant. Ten lessons each would have sufficed, but due to time limitations and other resource issues only three lessons were able to be given.

The study was deficient in resources, participants and time to determine whether Facebook could be of any use or of assistance to the elderly living in aged care accommodation, especially to combat social isolation. The results can be said to be inconclusive to confirm the hypothesis. This is due to the participants not being able to access the technology for a prolonged period of time, and not feeling confident enough to access technology on their own. However the personal encounters and experiences of the participants confirm that the computer lessons including the use of Facebook and email have been extremely useful and thoroughly enjoyable for all of the participants involved.

The research found common issues and problems experienced by all of the participants in the study:

1. the small print on the computer screen made it difficult to read the text;
2. using the mouse such as clicking and double clicking were made difficult due to the participants having shaking hands;
3. remembering user names and passwords was difficult at times, and
4. remembering the steps involved in how to log into email and Facebook accounts.

These issues were all addressed and did not stop the participants' progress or their use of computers. The education provided to participants needs to be practical and provided for a suitable period of time.

Another important aspect which came forth was that assumptions cannot be made upon a participant's ability to use the technology based solely upon a participant's age. Many elderly people were familiar with the technology and Social Networking Sites due to the influence of their children and grandchildren, media, and self-education. The oldest participant was the most keen to learn and had exponentially positive attitudes towards learning a new technology.

The problems discovered in implementing social media into an aged care facility included gaining access to devices and the provision of wireless. Some participants had personal computers and private Wi-Fi in their rooms; which was provided by their immediate family. As older people who reside in aged care facility receive only a small portion of their pension (the remainder goes to the aged care provider to provide for their care) it would be very difficult for them to personally afford internet access. Hence the aged care facility should consider providing both personal computers and wireless for its residents.

An interesting factor about the aged care facility where the research took place was the form of isolation and loneliness experienced. The participants stated that at times they did feel isolated from their families, but more importantly they felt isolated due to the lack of English speaking residents at the facility, to whom they could relate and engage in friendships. The participants stated they felt 'out of place', alienated, and different from the other residents.

The second set of interview questions brought forth that participants were equally interested in the technology and satisfied because they gained a new skill. The participants were not given the opportunity to socialise on Facebook for a sufficient amount of time to draw a solid conclusion as to whether Facebook assisted in their feeling of social isolation and if Facebook encouraged and facilitated reciprocated communications.

This research demonstrates that future studies are necessary. Further residential aged care facilities should make a concerted effort with helping residents master this technology in a bid to encourage social interaction. Doing so will provide an interesting and useful experience for an elderly person who resides alone in an aged care facility.

## 5   Conclusion

The paper discussed the possibilities of the use of Facebook (and computers) as a possible enabler of communication to ease feelings of social isolation for the elderly living in an aged care facility. The research utilised two sets of interviews to qualitatively determine each participant's attitudes toward a new technology, feeling of social connectedness, and the impact the implementation of computers and Facebook had on the participants. Moreover this paper has given an in-depth and personal view of six residents and their experiences in dealing with this technology.

The paper was successful in the intent to discredit a viewpoint that an elderly person cannot or does not have the capability or understanding or will to use a computer, the internet or social media websites. Furthermore the paper enlightens us about the possibilities and potential usefulness of implementing social media applications into an aged care facility. The findings bring forth important needs and wants of the elderly; including aspects to consider when introducing computers and social media to an elderly group of people and what challenges need addressing. Although this paper is not a comprehensive study on the topic, the findings support the validity of applications in a nursing home setting. More studies are needed to measure the validity of social media as a tool to assist with social isolation.

The most isolated elderly, who would most greatly benefit from social media usage, are those living in isolated individual accommodation rather than in an aged care facility. Future research is necessary to examine that population, extending the results of this pilot study. Such research would require:

1. Each participant to be provided with their own computer and internet access;
2. The research team to individually travel to the home of each participant.

In future research, we will investigate this issue as well as examine whether the administration of medicines to the elderly can be monitored via the use of social media. Based on pilot results that indicated difficulty in mastering the PC/Keyboard/Mouse interface, future research should consider using iPads/tablets instead of computers which may eliminate some of those barriers.

## References

1. Levy Cushman, J., McBride, A., Abeles, N.: Anxiety and depression: implications for older adults. J. Soc. Distress Homeless **8**, 139–156 (1999)
2. Saunders, E.J.: Maximising computer use among the elderly in rural senior centres. Educ. Gerontol. **30**, 573–585 (2004)
3. Dickens, A.P., Richards, S.H., Greaves, C.J., Campbell, J.L.: Interventions targeting social isolation in older people: a systematic review. BMC Public Health **11**(647), 1–22 (2011)
4. Department of Economic and Social Affairs: Population Division, World Population Ageing 1950–2050, United Nations, New York (2001). http://www.un.org/esa/population/publications/worldageing19502050/pdf/8chapteri.pdf. 20 Jan 2016
5. Australian Bureau of Statistics: 3222.0 Population Projections, Australia, 2012 (base) to 2101, Australian Bureau of Statistics (2013). http://www.abs.gov.au/ausstats/abs@.nsf/Lookup/3222.0main+features52012%20%28base%29%20to%202101. 20 Jan 2016
6. Feist, H., Parker, K., Howard, N., Hugo, G.: New technologies: their potential role in linking rural older people to community. Int. J. Emerg. Technol. Soc. **8**(2), 68–84 (2010)
7. National Ageing Research Institute (NARI) and La Trobe University: What Older People Want? Outcomes from a consultation with older consumers about their priorities for research in ageing. Consultation report 2012–2013 (2012–2013). http://www.nari.net.au/files/aaa_consultation_report_14_jan_2014.pdf. 21 Nov 2016
8. Seeman, T.E.: Health promoting effects of friends and family on health outcomes in older adults. Am. J. Health Promot. **14**(6), 362–370 (2000)
9. Mickus, M.A., Luz, C.C.: Televisits: sustaining long distance family relationships among institutionalized elders through technology. Aging Ment. Health **6**(4), 387–396 (2002)

10. Savenstedt, S., Brulin, C., Sandman, P.O.: Family members narrated experiences of communicating via video-phone with patients with dementia staying at a nursing home. J. Telemed. Telecare **9**, 216–220 (2003)
11. Dalgard, O.S., Håheim, L.L.: Psychosocial risk factors and mortality: a prospective study with special focus on social support, social participation, and locus of control in Norway. J. Epidemiol. Commun. Health **52**(8), 476–481 (1998)
12. Egan, N.: Seniors taking to the internet to reduce loneliness, Technology Review, 26 June 2015. http://www.australianageingagenda.com.au/2015/06/26/seniors-taking-to-the-internet-to-reduce-loneliness/. 28 Aug 2015
13. Carstensen, L.L., Fremouw, W.J.: The influence of anxiety and mental status on social isolation among the elderly in nursing homes. Behav. Resid. Treat. **3**(1), 64–80 (1988)
14. Oliver, D.P., Demiris, G., Hensel, B.: A promising technology to reduce social isolation of nursing home residents. J. Nurs. Care Qual. **21**(4), 302–305 (2006)
15. Pinquart, M., Sorensen, S.: Influences on loneliness in older adults: a meta-analysis. Basic Appl. Soc. Psychol. **23**(4), 245–266 (2001)
16. Australian Bureau of Statistics: Older persons internet use, Household use of information technology, Australia 2012–2013. Australian Bureau of Statistics (2013). http://www.abs.gov.au/ausstats/abs@.nsf/Lookup/B10BDF0266D26389CA257C89000E3FD8?opendocument. 26 Apr 2014
17. Bell, C., Fausset, C., Farmer, S., Nguyen, J., Harley, L., Fain, W.B.: Examining social media use among older adults. In: Proceedings of the 24th ACM Conference on Hypertext and Social Media, pp. 158–163. ACM, May 2013
18. Hope, A., Schwaba, T., Piper, A.M.: Understanding digital and material social communications for older adults. In: Proceedings of the SIGCHI Conference on Human Factors in Computing Systems, pp. 3903–3912. ACM, April 2014
19. Hutto, C., Bell, C.: January. Social media gerontology: understanding social media usage among a unique and expanding community of users. In: 2014 47th Hawaii International Conference on System Sciences, pp. 1755–1764. IEEE (2014)
20. Sinha, V., Subramanian, K.S., Bhattacharya, S., Chaudhuri, K.: The contemporary framework on social media analytics as an emerging tool for behaviour informatics HR analytics and business process. J. Contemporary Management **17**(2), 65–84 (2012)
21. Graham, J.E., Rockwood, K., Beattie, B.L., Eastwood, R., Gauthier, S., Tuokko, H., McDowell, I.: Prevalence and severity of cognitive impairment with and without dementia in an elderly population. The Lancet **349**(9068), 1793–1796 (1997)

# How to Help a Pedagogical Team of a MOOC Identify the "Leader Learners"?

Sarra Bouzayane[1,3(✉)] and Inès Saad[1,2]

[1] MIS Laboratory, University of Picardie Jules Verne, 80039 Amiens, France
{sarra.bouzayane,ines.saad}@u-picardie.fr
[2] Amiens Business School, 80039 Amiens, France
[3] MIRACL Laboratory, Institute of Computer Science and Multimedia,
3021 Sfax, Tunisia

**Abstract.** This paper proposes a method for the identification of the "Leader Learners" within Massive Open Online Courses (MOOCs) in order to improve the support process. The "Leader Learners" are those who will be mobilized to animate the forum. Their role is to help the other learners find the information they need during the MOOC so as not to drop it. This method is based on the Dominance-based Rough Set Approach (DRSA) to infer a preference model generating a set of decision rules. The DRSA relies on the expertise of the human decision makers, who are in our case the pedagogical team, to make a multicriteria decision based on their preferences. This decision concerns the classification of learners either in the decision class Cl1 of the "Non Leader Learners" or in the decision class Cl2 of the "Leader Learners". Our method is validated on a French MOOC proposed by a Business School in France.

**Keywords:** DRSA · Massive open online courses · Pedagogical team · Preference model · "Leader Learner"

## 1 Introduction

A MOOC is a form of online social learning taking into account the changing affordances of a world in which social activity increasingly takes place at a distance and in mediated forms [5]. According to Downes [8], the MOOC must not only be a simple transmission or use of information, but rather a development of a set of interactive activities and skills. Initially, MOOCs were proposed for an academic purpose only. But recently, they are being integrated into the organizations. In fact, according to a survey made by Future Workplace[1] and completed by 195 corporate learning and human resource professionals, 70% of the respondents are for the integration of MOOCs into their own companies learning programs. Other statistics[2] show that 34% of companies already offer corporate MOOCs for some of their employees, 32% of companies plan to introduce them by 2016 and 34% of companies do not plan for any

---

[1] http://futureworkplace.com/.
[2] http://blogs.speexx.com/blog/companies-already-use-corporate-moocs/.

© Springer International Publishing AG 2017
D. Bajwa et al. (Eds.): GDN 2016, LNBIP 274, pp. 140–151, 2017.
DOI: 10.1007/978-3-319-52624-9_11

MOOC integration. However, MOOCs still suffer from a high dropout rate that reaches the 90% [14]. Generally, a dropout action is linked to the difficulty for the pedagogical team to lead a better support process. In this context and according to a report prepared by Caron et al. [6] on a French MOOC broadcasted on the platform FUN[3], 32.6% of the learners were dissatisfied with the pedagogical team's support and 16.9% of them were very dissatisfied. Added to that, among the learners who requested assistance, 25% were dissatisfied with the answers they have received. Moreover, our study, on a French MOOC proposed by a Business school in France, revealed that more than 39% of the questions asked by the learners on the forum have been neglected, only 20% of them were considered by the pedagogical team and about 41% were handled by the learners themselves. In effect, a MOOC is characterized not only by a huge number of learners communicating remotely and having different cultural backgrounds, but also characterized by a growing mass of data coming from heterogeneous sources. These factors make it difficult for the pedagogical team to manage a satisfactory support process. According to Gooderham [9], it is the geographical and cultural distances between the transmitter and the receiver of the information that complicate the information transfer process and consequently the whole support process.

In this paper, our objective is, thus, to reduce the dropout rate through the identification of the "Leader Learners" in a MOOC. A "Leader Learner" is a learner having the required skills to assist the other learners when they need help, at any time during their participation in the MOOC. She/he will be mobilized to animate the MOOC forum in order to strengthen the pedagogical team. To this end, we propose a method based on the DRSA approach [10]. This method allows to classify learners either as "Non Leader Learners" belonging to the decision class Cl1 or as "Leader Learners" belonging to the decision class Cl2. The characterization phase is based on four steps: First, the cleansing of the initial data set of learners. Second, the construction of a family of criteria to characterize a "Leader Learner", based on the expertise of the pedagogical team. Third, the construction of a learning set of "Learners of Reference" using the 10-Fold Cross-Validation technique. Fourth, the inference of a preference model generating a set of decision rules using the algorithm DOMLEM proposed by the DRSA approach. This method is validated on a French MOOC proposed by a Business School in France.

The paper is structured as follows: Sect. 2 presents the related work. Section 3 details the approach DRSA. Section 4 presents the method of "Leader Learners" identification. Section 5 presents a case study. Section 6 concludes our paper.

## 2 Related Work

In literature, many works propose the models of prediction of the learners who plan to drop the learning environment. The work of Wolf et al. [18] associated the high dropout rate with the absence of face-to-face interaction. Wolf et al. proposed a model of prediction of the learners who would have a poor rating, and who are called

---

[3] https://www.fun-mooc.fr/.

"At-risk" learners. They combined the attribute of the number of clicks made by a learner on the learning environment with his demographics data and his access frequency to the evaluations space. Thus, a reduction of the number of clicks between two evaluation periods means that there is a strong probability that this learner will drop out from the training. The solution proposed is a telephone intervention to motivate the "At-risk" learner to continue his training. Viberg and Messina Dahlberg [17] linked the dropout rate to a commensurability problem. It is the case when the meaning given by the pedagogical team is not that understood by the learner. Besides, Kizilcec et al. [11] proposed a clustering method based on the completeness of the responses made by the learners on the proposed activities. This method resulted in four clusters. The cluster "Completing" groups learners having achieved the majority of the activities offered in the MOOC. The cluster "Auditing" represents the learners who rarely make evaluations but who remain engaged in the MOOC. The cluster "Disengaging" gathers the learners who responded to the MOOC activities only at its beginning. The last cluster is that of "Sampling" and contains learners having watched the video conference for only one or two evaluation periods. Finally, we cite Barak [1] who proposed a classification that is mainly based on the motivation of the learners about their involvement in the MOOC. In this context, the author has divided the students into three classes: the "Random visitors" who participate just to discover the MOOC, the "Novice students" who are beginner learners having a certain experience in distance education and who may drop the MOOC in a more advanced period, and finally, the "Expert Students" who are enrolled in the MOOC with the firm conviction to carry it on until the end so as to deepen their knowledge. The previous works have proposed methods to predict the dropout behavior or to characterize classes or clusters of learners on the basis of the identified behaviors. However, no work has proposed a concrete solution to improve the support process in order to reduce the dropout rate. Such a missing solution would allow the pedagogical team to manage the MOOC and the learners in order to improve their learning process. Thus, in this paper we propose a method of characterization of the "Leader Learners" to strengthen the pedagogical team in a context of MOOCs in order to minimize the dropout rate.

## 3    Dominance-Based Rough Set Approach

The approach DRSA was proposed by Greco et al. [10] and inspired from the Rough Sets Theory [13]. It allows comparing actions through a dominance relation and takes into account the preferences of a decision maker to extract a preference model resulting in a set of decision rules. According to DRSA, a data table is a 4-tuple $S = <K, F, V, f>$, where K is a finite set of reference actions, F is a finite set of criteria, $V = \cup_{g \in F} V_g$ is the set of possible values of criteria and $f$ denotes an information function $f: K \times F \rightarrow V$ such that $f(x, g) \in V_g, \forall x \in K, \forall g \in F$.

F is often divided into a subset $C \neq \emptyset$ of condition attributes and a subset $D \neq \emptyset$ of decision attributes such that $C \cup D = F$, and $C \cap D = \emptyset$. In this case, S is called a decision table. In multicriteria decision making, the scale of condition attributes should be ordered according to decreasing or increasing preference of a decision maker. Such attributes are called criteria. We also assume that the decision attribute set $D = \{d\}$ is a singleton. The unique decision attribute d partitions K into a finite number of decision classes $Cl = \{Clt; t \in T\}$, $T = \{1..n\}$, such that each $x \in K$ belongs to one and only one class. Furthermore, we suppose that the classes are preference-ordered, i.e., for all $r, s \in T$ such that $r > s$, actions from $Clr$ are preferred to actions from $Cls$. Among the parameters that the approach DRSA defines, we cite:

**Definition 1** (Decision rule) — Decision rules used in this paper are represented as follows:

$$\text{if } f(x, g_1) \geq r_1 \wedge .. \wedge f(x, g_n) \geq r_n \text{ then } x \in Clt^{\geq} \text{ such that } (r_1...r_n) \in (v_{g1}...v_{gn})$$

**Definition 2** (Strength or force) — We calculate the strength of a decision rule by the ratio of the number of actions supporting this rule and the number of all actions which were correctly classified in a given class.

**Definition 3** (Quality of approximation) — The following ratio measures the classification quality of a partition Cl using criteria set $P \subseteq F$. It expresses the percentage of actions that are assigned with certitude in a given class.

In this work, the actions are the learners enrolled in the MOOC and the Decision Makers are the pedagogical team of this same MOOC. Also, we consider only two decision classes: Cl1 of the "Non Leader Learners" and Cl2 of the "Leader Learners".

## 4 Preference Model Construction for the "Leader Learners" Characterization in a MOOC

In this section, we present a method based on the DRSA approach to construct a preference model resulting in a set of decision rules. This method allows to characterize the "Leader Learners" within a MOOC. Obviously, the pedagogical team of a MOOC consists of a set of persons who do not necessarily share the same preferences or the same decisions about the classification of each learner. Thus, to deal with such a problem of group decision, we would adopt the constructive approach of Belton and Pictet [2] that identifies three generic frameworks: (i) sharing—the group behaves as a single decision-maker and agrees on one common preference. It adapts the common and shared output of a discussion made between the different stakeholders; (ii) aggregating—the preferences of different stakeholders are aggregated and a common preference is obtained through voting or calculation. It uses a mechanistic approach to find consensus and requires the acceptance of the aggregate function by the group; and (iii) comparing—the stakeholders state their individual preferences and these are used in a negotiation process where the aim is to find a consensus. In this work, we adopt the first procedure based on sharing. In fact, after the detection of a conflict, we organize direct meetings with the members of the pedagogical team to identify the cause of the inconsistency and to reach then, a compromise. The method we propose consists of four steps:

## 4.1    Step 1: Cleansing the Initial Data Set of Learners

This step aims to remove redundant and incomplete information from the initial set of data about all the enrolled learners. In effect, learners can repeatedly enter their data to ensure that their registration is taken into account. In the case of MOOCs where we are faced with a growing body of data, removing redundancy is highly recommended. And since learners can interrupt the registration phase, they have to be neglected because of that missing data that would prevent us from characterizing them.

## 4.2    Step 2: Construction of a Family of Criteria

In this step, we used a constructive approach based on a literary review on the purpose of constructing a criteria family leading to the characterization of some learners as "Leader Learners". Morris et al. [12] showed that the experience a learner acquires on the e-learning and his study level are attributes permitting to predict his dropout rate in a MOOC. The study of Barak [1] proved that the MOOC language mastery is a factor motivating the learner to carry it on. This author stressed also the importance of the learner's motivation in relation to his participation in the MOOC. Likewise, Suriel and Atwater [16] emphasized the importance of the cultural background in guaranteeing a successful learning process to the learner. Added to the static data, experts also consider the dynamic data which are traced according to the learner's activity on the learning environment. Authors in [18] distinguished three types of activity that would permit us to predict the dropout of a learner. These are the access to a resource or to a course material; the publishing of a message on the forum and the access to the evaluation space. However, based on a thorough study on the domain of knowledge transfer, we define the dynamic criteria according to three dimensions: the sharability [4], the autonomy and the absorptive capacity [7].

– Sharability: this criterion is an attitude expressing the willingness of a transmitter to share his knowledge with the other members of his work unit [4]. In a MOOC context, a "Leader Learner" must show a high sharability showed through the number of messages he publishes weekly on the forum proposed by the MOOC. We define three types of messages:
  • Message of type "Add": it is a message posted on the forum and containing information that has not been requested by another learner.
  • Message of type "Response": this is a message added to answer a question already asked on the forum.
  • Message of type "Question": this is a message published by a learner in need of information.

We have considered the weekly number of each type of message as a criterion, on which the decision maker can rely to classify a learner as "Leader Learner" or "Non Leader Learner". The interaction on forums when participating in a MOOC training is a relevant indicator of the learners' sharability.

– Autonomy: this criterion expresses the learner's ability to have and find the information he needs. In this context, we measure the autonomy by the navigation frequency on the environment of the MOOC. It is calculated by the weekly number of viewed and/or downloaded resources (supports), as follows:

$$\text{Learner's autonomy} = \sum_{i=1}^{n} nb_{Consultations}(Support_i) \qquad (1)$$

Where $n$ is the number of the consulted supports and $nb_{Consultations}(Support_i)$ is the number of consultations of a given support $i$ by week.

– Absorptive capacity: this criterion is defined in the knowledge transfer field. It represents the ability of the learner to absorb the information he was provided with. This ability is demonstrated in a MOOC through the learner's answers on the activities proposed to him. The MOOC uses two types of activities: the multiple choice questionnaire (MCQ) and/or the peer-to-peer activity that is corrected by learners themselves. Generally, the MCQ fails to properly assess a learner. So, a great importance is granted to the peer-to-peer evaluation. The absorptive capacity or the score is calculated as follows:
  - If the week was closed with only a MCQ or a Peer-To-Peer activity, the score would be equal to the obtained mark.
  - If during the week, both activities are proposed, the score will be calculated as following:

$$\text{Score} = 0.2 * \text{Note}_{MCQ} + 0.8 * \text{Note}_{Peer-To-Peer} \qquad (2)$$

The MCQs are automatically scored. The Peer-To-Peer activities are both made and corrected by learners themselves, according to a scale set by the tutor.

This family of criteria must be validated by the pedagogical team of the MOOC. This pedagogical team has to define a scale of preference for each criterion. The constructed criteria family is presented in Appendix A.

In the criteria family, there exist two categorizations which are the ordinal/cardinal criteria and the static/dynamic ones. And while the categorization static/dynamic is specific to the context of MOOCs and is imposed by its evolving nature over time, the ordinal/cardinal type is a choice provided by the approach DRSA to represent the preferences of the decision makers on each criterion. Yet, both categorizations are subjective, so remain linked to the pedagogical team's preferences when making the classification decision. The static evaluation streams from the information contained in the form provided by the platform hosting this MOOC and filled by the participants at their first registration. This information is provided only once, a thing which justifies its static nature. Dynamic data are supplied by the information system that permits to manage the whole MOOC.

### 4.3 Step 3: Construction of the "Learners of Reference" Set

The learning set of "Learners of Reference" is used to infer a preference model. Thus, the higher the quality of the "Learners of Reference" set is, the more efficient the preference model gets. In this context, to select a profitable sample of "Learners of

Reference", we have used the 10-Fold Cross-Validation technique [15] which divides the original sample into 10 sub samples. Then, each of the 10 samples is selected as a set of validation and the 9 other samples as training sets. The operation is repeated k times until each sub-sample is used exactly once as a validation set. This technique is applied on a decision table that has to be completed by the pedagogical team. The lines of this table form the set of learners retained at the first step. The columns are the family of criteria constructed in the second step. The last column of this table contains the assignment decision of the leaner into one of the two decision classes: the decision class Cl1 of the "Non Leader Learners" or the decision class Cl2 of the "Leader Learners". The last column should be completed by the decision makers or pedagogical team during several meetings. In the decision making process, the decision makers are the individuals responsible for solving problems in order to attain a goal [3]. Thus, in order to dispel conflicts between the members of the pedagogical team we have used the constructive approach of Belton and Pictet [2]. The application of the 10-Fold Cross-Validation technique on this decision table provides ten training samples such that each one corresponds to a validation sample. Finally, on each training sample we have applied the DOMLEM algorithm proposed by the DRSA approach. This algorithm generates a minimal set of dominance–based rules covering all examples in the information table. At this level, we have not predefined requirement concerning, e.g., the minimal number of supporting actions or the maximal length of the condition part. However, if such a requirement exists we can use the AllRules algorithm or the DomExplore one, which are obviously characterized by a high computational cost and an exponential complexity with respect to a number of attributes, compared to the DOMLEM algorithm.

### 4.4    Step 4: Construction of a Preference Model

To select a preference model, we applied each of the ten preferences models previously inferred on the corresponding validation samples. Then, we compared the classifications given by each preference model with the real classifications made by the pedagogical team. The comparison is based on four measures:

- True positive (TP): the "Leader Learners" classified by the model as "Leader Learners".
- True negative (TN): the "Non Leader Learners" classified by the model as "Non Leader Learners".
- False positive (FP): the "Non Leader Learners" classified by the model as "Leader Learners".
- False negative (FN): the "Leader Learners" classified by the model as "Non Leader Learners".

Then, we calculated the precision and the recall measures. The precision reflects the number of learners correctly predicted by the preference model; the recall reflects the number of correctly predicted learners related to the positive examples and the F-measure represents the harmonic average of precision and recall.

$$\text{Precision} = TP/(TP + FP); \text{Recall} = TP/(TP + FN) \qquad (3)$$

$$\text{F} - \text{measure} = (2 * Precision * Recall)/(Precision + Recall) \qquad (4)$$

The preference model that must be selected is the one whose F-measure is maximal. Compared to some other methods and techniques that can be applied to cope with this problem, such as the artificial neural networks or the simple multivariate statistical clustering procedures, the DRSA approach gives a more sophisticated quality of learners' classification since it is based on the intervention of human decision makers for decision making process. Moreover, this approach permits to put out the inconsistencies and the missing data, if existing, in the decision table. Finally, the resulting preference model is presented in the form of decision rules of the type "if condition (s), then decision" and allows the decision maker to understand the reason behind his decision in a natural language.

In this paper, inferred decision rules aim at identifying among the massive number of learners those who have the ability to become leaders. As already said, these leaders will be mobilized by the pedagogical team to help them animate the forum. In this context, several scenarios may exist. For example, these leaders can just be contacted by the pedagogical team to motivate them to answer the forum posts. These leaders can be also identified by "caps" in order to be directly contacted by the learners in need. As it is the case for the OpenClassrooms[4] platform that identifies a set of "specialist learners" in the subject of the MOOC, to answer the questions asked by the other learners. These identified learners are characterized by what we call "caps". These are learners who are previously certified by the MOOCs offered by the same platform. Moreover, a recommender system can be proposed on the purpose of recommending to each learner in need, the set of leaders appropriate to his profile and that would support him during his MOOC training. However, our work is limited to identify these leader learners, but the point how they will be mobilized is not yet treated.

## 5    Case Study

This method is applied on real data collected from a French MOOC proposed by a Business School in France. This MOOC lasted five weeks. Our method of identifying "Leader Learners" was run as follows:

- Step 1: Following the MOOC, we obtained a Comma-Separated Values (CSV) File containing data about 2565 learners. We identified 15 redundant lines and 1030 learners with missing data. Therefore, the retained file contained data about only1520 learners.
- Step 2: To construct a family of criteria, we held several meetings with the pedagogical team in order to consider their viewpoint about the importance of each criterion. During these meetings, the pedagogical team members intervened to elicit their preferences concerning each criterion. The output of this step is presented in the Appendix A.

---

[4] https://openclassrooms.com/.

- Step 3: The pedagogical team fills the data in column D of the decision table (see Table 1). Such a column shows the classification of each learner either in the decision class Cl1 as "Non Leader Learner" or in the decision class Cl2 as "Leader Learner". Then we applied the 10-Fold Cross-Validation on this decision table. We obtained ten learning samples and ten validation ones. Finally, on each learning sample, we applied the algorithm DOMLEM. We ultimately obtained ten preference models.

**Table 1.** An extract from the decision table.

|    | $g_1$ | $g_2$ | $g_3$ | $g_4$ | $g_5$ | $g_6$ | $g_7$ | $g_8$ | $g_9$ | $g_{10}$ | $g_{11}$ | $g_{12}$ | $g_{13}$ | D |
|----|----|----|----|----|----|----|----|----|----|----|----|----|----|----|
| $L_1$ | 3 | 3 | 1 | 1 | 1 | 0 | 2 | 3 | 0 | 0 | 1 | 30 | 10 | Cl1 |
| $L_2$ | 2 | 2 | 3 | 2 | 1 | 0 | 4 | 2 | 1 | 0 | 0 | 44 | 10 | Cl2 |

- Step 4: To select the most sophisticated among the ten preference models previously inferred, we have calculated the precision, the recall and the F-measure of each model. The obtained measures are given in Fig. 1.

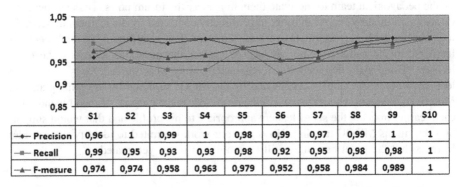

**Fig. 1.** Evaluation measures

Based on the F-measure results, we have selected the tenth preference model because it represents the maximal F-measure. An extract from the preference model is presented in Table 2.

**Table 2.** An extract from the preference model.

| Preference model based on the sample 10 | Force |
|---|---|
| **Rule 1. If** $g_9 \geq 1$ and $g_1 \geq 2$ and $g_{12} \geq 23$ and $g_3 \geq 3$ and $g_4 \geq 2$ and $g_8 \geq 2$ and $g_{13} \geq 8.5$ **Then** $L_i \in Cl_2$ | 21.05% |
| **Rule 2. If** $g_9 \geq 1$ and $g_2 \geq 3$ and $g_3 \geq 3$ and $g_{13} \geq 9.5$ and $g_1 \geq 2$ and $g_{12} \geq$ 17 and $g_7 \geq 3$ and $g_8 \geq 2$ **Then** $L_i \in Cl_2$ | 46.93% |
| **Rule 3. If** $g_9 \geq 1$ and $g_{12} \geq 35$ and $g_1 \geq 2$ and $g_3 \geq 3$ and $g_5 \geq 1$ **Then** $L_i \in Cl_2$ | 22.81% |
| **Rule 4. If** $g_1 \geq 2$ and $g_9 \geq 1$ and $g_{12} \geq 45$ and $g_{13} \geq 10$ **Then** $L_i \in Cl_2$ | 56.58% |

For example, the rule 4 can be translated as follows:

If the learner is "at least" a high school student and he added "at least" one message per week on the forum and he consulted "at least" 45 times per week the resources and he got "at least" 10 on the proposed activity of this week Then he is affected in the decision class "Leader Learners".

We note that the preference model selected is rich in criteria. In fact, all the criteria we identified were considered in at least one decision rule. In addition, the criterion $g_9$, indicating the weekly number of the messages added by a learner, was mentioned in all the decision rules of the model and that shows its importance. This result is logical since a "Leader Learner" is assigned with the role of sharing information on the forum. Thus, the number of messages that a learner has already published can be a relevant criterion in the process of characterization of the "Leader Learners". The quality of this preference model is also proved by the force of the decision rules it contains. We found two rules whose force exceeds 40% and two others with a force greater than 20%. This good quality may be caused by the human intervention in the decision making process. This highlights the efficiency of the approach DRSA.

When experimenting this method, we have considered only the decision rules whose strength is greater than 10%. However, the information about the number and the strength of rules supporting the classification of each "Leader Learner" are not yet used. In fact, despite the huge number of learners, the number of the identified leaders remains always limited compared to that of the learners in need. Consequently, we will never eliminate a leader even if he was identified on the basis of one small rule, because the higher the number of leaders is, the better the support process becomes.

This method can be applied on a similar MOOC (same pedagogical team, same MOOC subject, same MOOC complexity, etc.) or on a different one. In the second case, we must mobilize the new pedagogical team to adapt the criteria family to its members' preferences and to the features of the new MOOC.

## 6 Conclusion

To our best of knowledge, in literature, there are no works concerning the characterization of the "Leader Learners" despite the fact that this necessity was reported by some experts in this domain. Thus, in this paper, we proposed a method of four steps to identify the "Leader Learners" in a MOOC. A "Leader learner" is assigned with the role to animate the forum of a MOOC in order to strengthen the pedagogical team. This proposed method is based on the construction of a coherent family of criteria, the construction of a representative sample of "Learners of Reference" using the 10-fold cross-validation technique and the application of the approach DRSA to infer a preference model that would generate a set of decision rules. The DRSA approach relies on the experience of the human expert decision makers which would lead to a set of decision rules of good quality. As far as the future work is concerned, we plan to propose an approach allowing to automate the conflicts resolution between the members of the pedagogical team throughout the decision making process. Such an approach will be also useful if we want to reach a decision group between two or more pedagogical teams coming from different MOOCs. In this case the constructive approach of Belton and Pictet [2] is no longer valid because of the geographical distance that would separate the decision makers.

## A List of the Constructed Criteria Family

| Criterion | Description | Scale | P |
|---|---|---|---|
| $g_1$: Study level | Indicates the actual study level of the learner or the last diploma he obtained | 1: Scholar student; 2: High school student; 3: PhD Student; 4: Doctor | ↑ |
| $g_2$: Level of technical skills | Indicates the extent to which the learner masters the use of the computer tools | 1: Basic; 2: Average; 3: Expert | ↑ |
| $g_3$: Level of proficiency in MOOC language | Indicates the extent to which the learner masters the language of the MOOC | 1: Basic; 2: Average; 3: Good | ↑ |
| $g_4$: Motivation for MOOC registration | Indicates the motivation behind the participation of the learner in the MOOC | 1: Just to discover the MOOCs; 2: To exchange ideas with the other learners or to have a certificate; 3: To exchange ideas with the other learners and to have a certificate | ↑ |
| $g_5$: Previous experience with MOOCs | Indicates whether the learner has a previous experience on learning via MOOCs or not | 0: No experience at all; 1: At least one experience | ↑ |
| $g_6$: Mastery level of the subject of the MOOC | Indicates to which extent the learner masters both the topic and the theme of the MOOC | 0: No knowledge at all; 1: Average knowledge; 2: Deepened knowledge | ↑ |
| $g_7$: Probability to finish the MOOC | Indicates the probability for a learner to carry-on the MOOC activities until the end | 1: Very weak; 2: Weak; 3: Average; 4: Strong; 5: Very strong | ↑ |
| $g_8$: Weekly availability predicted | Indicates the estimative weekly availability of the learner to follow the MOOC | 1: Less than one hour; 2: From one to two hours; 3: From two to three hours; 4: Four hours or more | ↑ |
| $g_9$: Number of add message | Indicates the number of the messages added on the forums per week | $n \in n$; $n \geq 0$ is the maximum number of the added messages per week | ↑ |
| $g_{10}$: Number of responses published on the forum | Indicates the weekly number of the responses to an asked question published on the forum | $m \in n$; $m \geq 0$ is the maximum number of answers per week | ↑ |
| $g_{11}$: Number of questions asked on the forum | Indicates the weekly number of questions asked by learners on the forum | $k \in n$; $k \geq 0$ is the maximum weekly number of questions | ↑ |
| $g_{12}$: Frequency of navigation on the MOOC site | Indicates the capacity of the learner to interact with the site. It is calculated upon the number of resources consulted by week | $p \in n$ such that $p \geq 0$ is the weekly number of site consultation by the learner | ↑ |
| $g_{13}$: Score | Indicates the weekly score the learner got on the set of activities he made | The note $\in [0, 10]$ | ↑ |

# References

1. Barak, M.: The same mooc delivered in two languages: examining knowledge construction and motivation to learn. In: Proceedings of the EMOOCS, pp. 217–223 (2015)
2. Belton, V., Pictet, J.: A framework for group decision using a MCDA model: sharing, aggregation or comparing individual information. Revue des Systemes de Decision **6**, 283–303 (1997)
3. Bolloju, N., Khalifa, M., Turban, E.: Integrating knowledge management into enterprise environments for the next generation decision support. Decis. Support Syst. **33**, 163–176 (2002)
4. Boughzala, I., Briggs, R.O.: A value frequency model of knowledge sharing: an exploratory study on knowledge sharability in cross organizational collaboration. Electron. Mark. **22**(1), 9–19 (2012)
5. Buckingham Shum, S., Ferguson, R.: Social learning analytics. Educ. Technol. Soc. **15**(3), 3–26 (2012)
6. Caron, P.A., Heutte, J., Rosselle, M.: Rapport d'Expertise et Accompagnement par la recherche du dispositif expérimental MOOC iNum (2012)
7. Cohen, W., Levinthal, D.: Absorptive capacity: A new perspective on learning and innovation. Adm. Sci. Q. **35**, 128–152 (1990)
8. Downes, S.: What Makes a MOOC Massive. Blog Post (2013)
9. Gooderham, P.: Enhancing knowledge transfer in multinational corporations: a dynamic capabilities driven model. Knowl. Manag. Res. Pract. **5**(1), 34–43 (2007). doi:10.1057/palgrave.kmrp.8500119
10. Greco, S., Matarazzo, S., Slowinski, S.: Rough sets theory for multicriteria decision analysis. Eur. J. Oper. Res. **129**, 1–47 (2001)
11. Kizilcec, R., Piech, C., Schneider, E.: Deconstructing disengagement: analyzing leaner subpopulations in massive open online courses. In: Third International Conference on Learning Analytics and Knowledge LAK 2013, pp. 170–179 (2013)
12. Morris, N., Hotchkiss, S., Swinnerton, B.: Can demographic information predict MOOC learner outcomes? In: Proceedings of the EMOOC Stakeholder Summit, pp. 199–207 (2015)
13. Pawlak, Z.: Rough sets. Int. J. Comput. Sci. **11**(5), 341–356 (1982)
14. Rayyan, S., Seaton, D.T., Belcher, J., Pritchard, D.E., Chuang, I.: Participation and performance in 8.02x electricity and magnetism: the first physics MOOC from MITx. arXiv. In: Proceedings of the Physics Education Research Conference (2013)
15. Refaeilzadeh, P., Tang, L., Liu, H.: Cross-validation. In: Liu, L., Özsu, M.T. (eds.) Encyclopedia of Database Systems, pp. 532–538. Springer, Heidelberg (2009)
16. Suriel, R.L., Atwater, M.M.: From the contributions to the action approach: white teacher. J. Res. Sci. Teach. **49**(10), 1271–1295 (2012)
17. Viberg, O., Messina Dahlberg, G.: MOOCs' structure and knowledge management. In: Proceeding of the 21st International Conference on Computers in Education, Depansar Bali, Indonesia (2013)
18. Wolff, A., Zdrahal, Z., Herrmannov, D., Kuzilek, J., Hlosta, M.: Developing predictive models for early detection of at-risk students on distance learning modules. In: Workshop: Machine Learning and Learning Analytics at LAK, Indianapolis (2014)

# Negotiating Peace: The Role of Procedural and Distributive Justice in Achieving Durable Peace

Daniel Druckman[1,2,3](✉) [iD] and Lynn Wagner[4]

[1] George Mason University, Fairfax, VA, USA
dandruckman@yahoo.com
[2] Macquarie University, Sydney, Australia
[3] University of Queensland, Brisbane, Australia
[4] International Institute for Sustainable Development, Winnipeg, Canada
lynnwagner@jhu.edu

**Abstract.** Many civil wars have been terminated with a peace agreement that ends the fighting, but these agreements have not always resulted in lasting peace. Earlier research on peace agreements has missed important points during which justice principles can play a role in establishing durable peace – during the negotiation process itself (procedural justice: PJ) and as incorporated into the negotiated outcome (distributive justice: DJ). Nor has the earlier research simultaneously considered the variety of dimensions that define durable peace, including reconciliation, security reform, governance, and economic growth. This study fills these gaps by examining the relationship between the justice and peace variables in 50 civil wars. Our analyses show that PJ and DJ led to more stable agreements and to a more durable peace: A significant time-lagged path from the justice to peace variables was demonstrated. The results suggest that just negotiation processes and outcomes are important contributors to peace.

**Keywords:** Distributive justice · Durable peace · Peace agreements · Peacekeeping · Procedural justice · Stable agreements

## 1 Introduction

As a peace and conflict analyst, you have just received a challenging assignment: Develop advice for the parties to a civil war on how to structure the negotiation of a peace agreement and what principles to incorporate into it. The parties previously reached a peace agreement that remained in force for several years, and was therefore touted as successful by some. This time around, they want advice on principles to incorporate into the negotiation process and outcome that could bring something more than a peace agreement that remains in force: How can they foster durable peace within the country?

This paper received the best paper award at the June 2016 Group Decision and Negotiation (GDN) conference in Bellingham, Washington.

The original version of this chapter has been revised: Copy editing mistakes have been corrected throughout the chapter. The erratum to this chapter is available at DOI: 10.1007/978-3-319-52624-9_13

© Springer International Publishing AG 2017
D. Bajwa et al. (Eds.): GDN 2016, LNBIP 274, pp. 152–174, 2017.
DOI: 10.1007/978-3-319-52624-9_12

Attaining durable peace after a civil war has proven to be a significant challenge, as many negotiated agreements lapse into violence. Since the end of World War II, civil wars have occurred in all corners of the world, with a dramatic increase in frequency following the end of the Cold War particularly on the African continent. (See the UN Peacemaker database, which contains close to 800 post-WWII cases that can be understood broadly as peace agreements.) Many civil wars have been terminated with a peace agreement that ends the fighting. Fewer agreements have resulted in successful implementation or a lasting peace between the warring parties.

Problems of implementing the terms of the agreements are evident in Downs and Stedman's [19] analyses of 16 peace agreements: Only 38% of their cases resulted in successful implementation. A larger sampling of 50 cases accumulated for the present research shows only a 34% success rate in reducing violence and a 16% success rate for peacekeeping (PK) missions. Similar results occur for achieving long-term peace, with success rates at 25% and 30% for the 16 and 50 case databases respectively. These data raise the question: Why do so many peace agreements fail to achieve their goal of bringing peace to war-torn nations? We address this question in this study, and in doing so build on the justice and negotiation literatures to explore how our analyst in the opening paragraph should respond.

Whether focusing attention on the short-term implementation of the provisions of agreements or on long-term improvements in relationships between former combatants, the success rates are disappointing. This lack of implementation may be due in large part to the conflict environment surrounding the talks. Indeed, the failed and partial success cases in the Downs and Stedman [19] dataset were negotiated in more difficult environments than the cases that were successfully implemented: The correlation between difficulty and implementation success was −.65 [21]. This correlation reduces somewhat when controlling for principles of justice, particularly the principle of equality, in the negotiated outcome. Previous studies found that these principles served to reduce the negative effects of the conflict environment on the short-term implementation of the agreements. Further research has shown that adhering to justice principles during the negotiation process and in the outcome may play important roles in addressing the challenges of implementation. We review this research below and develop further hypotheses, which we explore in a 50 case sample, to enhance our understanding of the relationship between justice and durable peace.

## 2   Justice in Negotiation

This study examines the impacts of procedural (PJ) and distributive (DJ) justice principles in negotiation on short and long-term peace. The PJ/DJ distinction is particularly relevant to negotiation, the one referring primarily to the way the process is conducted, the other to allocation decisions adopted at the conclusion of the negotiation. Procedural justice refers to principles for guiding the negotiation process toward agreements. These principles include fair treatment and fair play, fair representation, transparency, and voluntary decisions. One or more of these principles surface during the negotiation process either positively, as for example more fair play or transparency, or negatively, as for example a lack of fair play or transparency. Positive adherence to one or more of these principles usually move the process in the direction of agreement, whereas negative adherence often sustains impasses. Distributive justice refers to principles for allocating benefits or burdens among the members of a group or community. Four DJ principles are emphasized in the literature: equality, proportionality or equity, compensation, and need. One or more of these principles usually surface in the outcome of a negotiation.

Several studies showed strong effects for PJ. In their legal disputes simulations, Hollander-Blumoff and Tyler [28] found that more adherence to PJ principles led to more agreements and to more integrative agreements when they were available. In an archival analysis of historical peace agreements, Wagner and Druckman [42] showed that PJ led to more integrative outcomes when problem-solving processes were set in motion. In their analysis of 16 peace agreements, Albin and Druckman [2] demonstrated that PJ led to more durable outcomes when the principle of equality was central in the agreement texts. Based on the same 16 peace agreements, Wagner and Druckman [43] found a strong relationship between PJ and long-term reconciliation when the provisions of the agreements were not violated. Strong effects for PJ were also found in the arena of trade agreements: More effective outcomes were obtained when PJ principles were adhered to in both bilateral and multilateral trade talks [3]. These findings provide a foundation for this study.

With regard to DJ, a number of studies have shown that negotiators demonstrate preferences for one or another principle for dividing resources. The preference turns on the goal of the negotiation [29]. Equality is emphasized in situations that emphasize solidarity or coordination among group members [15]. Equity or proportionality is often the guiding principle in competitive performance situations [1, 23]. When indemnification for undue costs is salient, a compensatory principle is preferred [30]. And, when personal development or welfare is emphasized, negotiators invoke need as a guiding principle [11, 15]. Preference for one or another of these DJ principles in peace agreements may turn on the type of issue being discussed: equality is preferred as a way of dealing with reconciliation issues [2]; proportionality is preferred for economic issues [43].

Although the justice literature is divided between PJ [e.g., 31] and DJ [e.g., 15] studies, the concepts are related in several ways. Strong correlations between PJ and DJ were reported in the Hauenstein et al. [25] meta-analysis. These are, however, qualified by issue area, with variation in the strength of relationship in such domains as trade, arms control, and environmental negotiations. (See 22 for a summary of the range of correlations by issue area.) Statistical mediation effects are also shown to occur with regard to outcomes and to the durability of agreements: for example, the PJ-durability relationship is shown to be mediated by the DJ principle of equality [2]. A third type of relationship between PJ and DJ is in terms of compensatory effects, as when distributive losses are cushioned by fair procedures or when favorable distributive outcomes mollify the negative effects of unfair practices [7, 8]. These offsetting or sequential effects are referred to as compound justice [44]. But there are other ways of combining the principles, as noted by Cook and Hegtvedt [13], including a broader concept of perceived justice or fairness where both procedural and distributive elements are considered together. (See 25 for a discussion of the relevance of a global justice concept.)

Justice considerations influence the process and outcome including its durability in a variety of negotiation domains. The importance of PJ principles seems to turn on other features of the negotiation process (e.g., problem solving), goals, and issue area. The relevant distributive principle used by negotiators (e.g., equality or equity) seems to depend on issue area, goals, and task framing. Both types of justice are influenced as well by the larger environment in which negotiations occur [21]. Yet, it may be that the role of justice is understood best as a synergistic process emphasizing an interplay between PJ and DJ. Focusing on the implementation of peace agreements, we analyze both separate and combined impacts of justice principles on the stability of the agreement and durability of the peace following the agreements, as captured by the hypotheses presented below.

A growing body of literature has focused on the role of another type of justice. Transitional justice (TJ) addresses civil and political rights violations and provides redress or reparations for those who suffered as a result of the conflict [39, 40]. TJ plays an important role in securing a lasting peace. It kicks in usually following agreement on a peace accord and is relevant during the implementation stage. As such, it is conceived as part of the reconciliation component of our durable peace index. Our focus on the negotiation process and outcome leads us to concentrate on the impacts of PJ and DJ, both considered as independent variables, on peace processes following the agreement, considered as dependent variables. We turn now to a discussion of the peace indexes.

# 3  Stable Agreements and Durable Peace

A key question asked in this research is about the relationship between stable agree-ments (SA)[1] and durable peace (DP) in the context of civil wars. The former refers to the implementation of agreements to end civil wars and to violations of the provisions of these agreements [19, 24]. Stable agreements keep the conflict in check through cease fires and related provisions that manage a conflict. The latter concerns the extent to which peace is achieved in the war-torn post-conflict society and is assessed over a relatively long period of time. Durable peace occurs when the former combatants reconcile their differences and rebuild security, governmental, and economic institu-tions [36]. These activities are often part of the provisions of the agreements, but can be assessed independently, as variables that are components of durable peace rather than as peace agreement elements to be implemented. Distinguishing between conflict management and resolution [20], negotiated agreements can be evaluated in terms of their success in managing (as in stable agreements) and/or resolving (as in durable peace) conflicts between warring parties.

In their 16-case comparative study, Wagner and Druckman [43] showed a strong relationship between SA and DP. It appears that SA sets the stage for DP. Progress toward reconciliation and institution-building is made when parties adhere to the

---

[1] In our previous work, we used the phrase durable agreements (DA). In this paper, we use the phrase stable agreements (SA). The change in terminology was motivated by an attempt to clarify the distinction between implementing agreements and societal peace. Using the word "durable" for both may be confusing to readers. Thanks go to Ari Kacowicz for suggesting this distinction.

agreement's provisions and peacekeepers depart. The research also found indirect relationships between SA and DP. SA served as a mediating variable between the DJ principle of equality and DP as well as accounting for the relationship between PJ principles and reconciliation. Thus, SA serves to connect justice during the process and in the outcome with lasting peace. We explore these relationships further in this study with a larger sample of cases.

In the Wagner-Druckman study, SA was coded in terms of the three Downs and Stedman [19] implementation categories: success, partial success, and failure. These categories refer to whether large-scale violence is brought to an end while peace-keeping implementers are present and whether war is terminated on a self-enforcing basis so that implementers can leave without the war rekindling within a two-year period. Thus, two aspects of post-agreement implementation are included in the Downs and Stedman assessment: violence abatement and peacekeeping duration.

The DP index is original in our research. It consists of four components that capture the reconciliation and institutional features of post-agreement implementation. The reconciliation component overlaps to some extent with indicators used to assess transitional justice in the post-conflict period, such as Binningsbø et al.'s [6] post-conflict justice variables: trials, truth commissions, reparations, amnesties, purges and exiles. But our DP index brings these transitional justice variables together with additional elements highlighted in the literature as important for creating a lasting peace, such as security and legal reform as well as the extent to which sustainable economic growth has occurred. These indicators were monitored for eight years after the agreement was reached.

Recognizing that peace agreements are often implemented in difficult contexts, the research also incorporates a new assessment of the conflict environment in which the agreements are implemented. The previous evaluations included a variable related to the difficulty of the options available to leaders [19]. This research project adds an assessment of the toll that the violence has taken on the country's population. The new conflict environment and durable peace variables extend the earlier analyses from a focus on factors that influence stability of the agreement two years out to a focus on durable peace eight years after the agreement is reached.

The sections to follow are organized into several parts. First, we present the research that this study builds on and the hypotheses that are suggested by that research. Sampling methods are then described followed by the cases selected, coding procedures for each variable including calibrations, and statistical methods. Results on relationships among the coded variables are organized by the hypotheses. The final section develops implications of the results, suggests areas for further research, and presents conclusions.

## 4 Hypotheses

This section presents hypotheses intended to capture a variety of relationships among the peace and justice variables. A discussion of relevant prior research precedes each hypothesis. The hypotheses address the relationship between SA and DP and among the DP components, as well as the role of justice in SA and DP. The latter include mediating effects of the justice variables. A final pair of mediating hypotheses deals with the path that connects the two justice variables with the two peace variables.

## 4.1   Stable Agreements and Durable Peace

Conditions for SA, variations of which are often examined as the dependent variable in civil war and peace research, include violence abatement and containment. These objectives are often implemented by peacekeeping operations. Achieving these goals is an indicator of successful conflict management and a signpost to peacekeepers to exit their missions [17]. To the extent that these goals are attained, the agreement may be considered stable. Stable agreements may set the stage for durable peace. Durable peace is assessed not on the basis of the implementation of the peace agreement, but rather based on variables that have been found to contribute to lasting peace. The strong correlations and mediating relationships between SA and DP found in the Wagner and Druckman [43] study provide evidence for this pattern. This correlational relation, captured by the first hypothesis, is evaluated with a larger sample in this study.

*H1: Stable agreements correlate with durable peace. If a peace agreement has high (low) stability, then it will also have high (low) durable peace.*

As discussed above, the SA index used in the earlier study, based on the Downs and Stedman [19] implementation index, combines violence reduction with peace-keeper presence during a two-year period. These components are separated in this research. A strong correlation between them would suggest that they are similar or overlapping indicators of SA: A reduction in violence would co-vary with shorter PK missions as suggested by the following hypothesis:

*H2: Violence reduction is correlated with peacekeeping mission duration. If violence is reduced (increased), then peacekeeper missions will be of shorter (longer) duration.*

The DP index that we constructed for this research consists of four components: reconciliation, security institutions, governance, and economic stability or growth. Together, these components are an attempt to capture the DP concept. The earlier research showed that reconciliation and change in both security and governing institutions are highly correlated. These components of DP form a cluster of backward (reconciliation) and forward-looking (security and governance institutions) peace. An attempt is made to evaluate this finding with the larger sample in this study. The finding is a basis for the following hypothesis:

*H3: Reconciliation and institutional change are correlated. If reconciliation is achieved (not achieved), then security and governing institutions will be strengthened (not strengthened).*

In the Wagner and Druckman [43] study, the economic component of DP did not correlate with the other parts of the DP index. This finding suggests that economic benefits are largely independent of other changes that occur during the post-war period of recovery following a peace agreement. More broadly, it challenges arguments about economic growth as a dividend of peace [32]. It does however support Collier's [12] institutional argument and O'Reilly's [35] results on economic development during recovery from civil wars. Collier argued that institutional structures must be developed to encourage private investment needed for post-conflict economic recovery. Weak or uncertain institutions prevent investment from contributing to economic recovery. O'Reilly's analyses suggest that it takes at least six years following the dislocations caused by civil wars for institutional development to take place. The eight-year post-conflict period used to assess DP in this study may not provide the time needed for institutional development. This argument, together with the earlier findings, suggests the following hypotheses:

> H4a: *Economic growth does not occur during the short-term post-war recovery period.*
> H4b: *The economic component of DP is uncorrelated with reconciliation and the development of security and governing institutions.*

Support for these hypotheses has implications also for relationships between the justice variables and economic growth. These implications are considered further with the hypotheses to follow on the way justice influences DP.

## 4.2    Procedural and Distributive Justice

Research on PJ and DJ has developed primarily in the literatures of social and organizational psychology. Several studies have shown that fairness perceptions are enhanced in situations characterized by uncertainty, importance, severity, and scope or the size of the identity group being represented [9,10,37]. These features describe the situation faced by peace negotiators. Thus, justice perceptions are likely to play an important role in this domain as well, and this is demonstrated by several recent studies. Building on these studies, we investigate various ways in which justice effects may occur.

Of particular interest are the debates in these literatures about the relationship between the types of justice. Four issues are discussed: relative effects of the two types of justice, separate and combined effects, interactive effects, and mediated effects. With regard to relative effects, we ask whether PJ and DJ impact on different aspects of SA and DP? Earlier research showed strong effects of DJ, particularly the equality principle, on durability. It also showed that the proportionality principle correlated strongly with the economic component of DP. PJ on the other hand was related strongly to reconciliation. The distinction suggested by these findings is between economic and socio-emotional motives: DJ concerns the distribution of resources; PJ is related more closely to the development of a relationship between negotiators. Ambrose and Cropanzano [4] suggested that PJ and DJ perceptions derive from expectations, which may at times be economic and at other times socioemotional. Thus, three hypotheses are suggested:

> H5: *The more central is the equality principle in the agreement, the more stable is that agreement.*
> H6: *The more central is the proportional principle in the agreement, the stronger is the economic growth in the society.*
> H7: *The more that parties adhere to PJ principles during the process, the stronger is the reconciliation between them during the post-agreement period.*

The mediating role played by equality is captured by the following hypothesis.

*H8: When the equality principle is emphasized in the agreement provisions, adherence to PJ principles lead to more stable agreements.*

The mediating role played by stable agreements is captured by the following two hypotheses:

*H9a: An emphasis on the equality principle leads to durable peace when the agreements are stable.*
*H9b: Adherence to PJ principles during the process leads to reconciliation when the agreements are stable.*

The longer path, with two mediating variables, has not been explored in earlier research. That path consists of a progression from PJ to DJ to SA to reconciliation or DP: Adherence to PJ principles during the process encourages adherence to DJ principles in the agreement, which, in turn, leads to more stable agreements and reconciliation or durable peace. The progression, referred to as serial mediation, is summarized by two hypotheses, one (H9c) that includes reconciliation as the dependent variable, the other (H9d) including DP as the dependent variable:

*H9c: Adherence to PJ principles leads to reconciliation when negotiators adhere to DJ principles and the agreements are stable.*
*H9d: Adherence to PJ principles leads to durable peace when negotiators adhere to DJ principles and the agreements are stable.*

This set of hypotheses is evaluated with statistical analyses to be described in the next section.

# 5   Cases, Variables, and Analyses

This study draws on and contributes to research in the scientific study of international processes tradition [34]. We code each case according to variables that are identified by our hypotheses, and conduct statistical analyses to discover relations among them. Our coding draws on datasets that peace agreement scholars often use – the Uppsala Conflict Data Program (UCDP), the Peace Accords Matrix (PAM) and the UN Peacemaker database. However, because our hypotheses require the coding of new independent (justice) and dependent (durable peace) variables, we chose to focus on a subset of the cases. Fifty cases were randomly selected to represent the universe of available cases in these datasets. To further ensure the validity of the findings, we develop conceptually distinct definitions of each variable and use different material for coding the variables. This section describes the sampling and coding procedures as well as the calibration methods that were used to develop the data set.

## 5.1   Sampling Procedures

A random sampling procedure was used to ensure that both partial and comprehensive outcomes were represented in this research, as well as agreements struck during a variety of decades. The cases range from the Bandranayaki-Chelvanayakam Pact in Sri Lanka, which was signed in 1957, to the Djibouti Agreement in Somalia, which was concluded in 2008. Given that cases in Africa are overrepresented in more recent decades, the selection by decade also ensured a more representative sampling of global regions.

Three datasets established the universe of possible peace agreement cases for this study. The PAM database consists of 35 civil war cases, all of which achieved "comprehensive" agreements. Because these cases arguably offer the greatest insights into how to do a peace agreement "right," we randomly selected 25 of the PAM agreements for this study. We looked to the UCDP and UN Peacemaker datasets to identify partial agreements and ensure variety among the decades. Of the UCDP database's 217 comprehensive and partial peace agreements that were signed between at least two opposing primary warring parties in an armed conflict between 1975 and 2011, 28 civil war cases were partial agreements that were not signed by all warring parties, with 17 signed in the 1990s and 11 in the early 2000s. We randomly selected nine cases from the 1990s and six from the 2000s, based on the ratio of cases during the 1990s to 2000s. The UN Peacemaker database includes close to 800 documents that can be understood broadly as peace agreements, including 44 intra-State peace agreements that had been signed prior to 1990: 34 during the 1980s and 10 prior to 1980. We randomly selected seven cases from the 1980s and three from earlier decades, based on the ratio of cases during the 1980s to earlier decades. Supplementary Appendix I lists the 50 peace agreements examined in this research. The Appendix is available from the authors.

## 5.2   Coding Procedures

This research builds on the peace agreement literature, but extends it in several ways. For example, Hartzell and Hoddie [24], Martin [33] and others examine the relationship between power sharing in peace agreement "outcomes" and the "durability" of peace, as measured by the number of years that an agreement remains in force. Power sharing is one component (equality) of distributive justice, but this research project incorporates additional justice aspects that could be incorporated into an outcome, as identified in the negotiation and justice literature (see 22). We build on research that finds a relationship between procedural and distributive justice and stable agreements [2,21], using a similar coding procedure for PJ and DJ. These studies relied on a measure of SA that Downs and Stedman [19] developed, which incorporated measures of the end of violence with assessments of whether peacekeeping forces were able to leave.

Our research departs from many other peace agreement studies in its measurement of the dependent variables: we assess whether violence ended and if peacekeepers were able to leave, rather than the number of years that an agreement remained in force. We also develop a new index for positive, durable peace (DP), which assesses the extent of reconciliation and institution building in the areas of security, governance, and economics for an eight-year period after the agreement was signed. This DP variable is coded independently of the negotiation outcome and its stability. The following sections present the coding procedures used for each of the variables: procedural justice, distributive justice, stable agreements, and durable peace. These procedures are summarized in Supplementary Appendix II available by writing to the authors.

**Procedural Justice.** Primary and secondary accounts of each negotiation process were evaluated to identify instances in which four procedural justice elements played a role in the negotiations: transparency, fair representation, fair play/fair treatment and voluntary agreement. If any of these PJ variables played a partially or fully satisfied role in the negotiation process, the variable was assessed for whether it had a "highly significant," "important" or "marginal" influence on the process. Satisfied variables were those instances in which the justice component had a positive influence on the talks: relevant information was shared, stakeholders were included in the talks and had the opportunity to make their voice heard, and the outcome was agreed to without coercive tactics.

Highly significant instances of PJ were assigned "3," important instances of PJ were assigned "2," and marginal instances of PJ were assigned "1." For each of the four PJ variables, we add the relevant scores and develop an average of the scores for that PJ variable. We then add these averages per PJ variable to develop a "PJ sat" assessment. A student in the graduate program at Uppsala University was recruited as a research assistant on the project to code each case according to guidelines established for this study. The principal investigators reviewed the data for 19 randomly selected cases and developed independent coding assessments, which matched 87.8% of the primary coder's assessments. The calibration procedure concluded with a discussion about those codes that differed, and adjustments to the coding were made.

**Distributive Justice.** A second research assistant assessed the provisions in the signed peace agreements for each of the 50 cases according to four principles: equality, proportionality, compensatory justice, and need. Like PJ, DJ was judged in terms of its significance, or centrality, to the agreement. Each article in each peace agreement was assessed for whether it represented one of the four DJ principles. The coder then considered whether all of the elements representing each principle resulted in the principle playing a highly significant (1.5), very important (1.25), important (1), less important (.75), or marginal (.5) role in the outcome. The intervals between the highly significant, important and marginal codes are proportional to the intervals between the codes for the comparable PJ measures. However, our practice coding with the DJ measure suggested that we could make finer distinctions. Thus, two additional categories were added to the coding scheme. The coder made these assessments based on whether the agreement would have been fundamentally different without the elements that embodied that principle. The numerical values for each principle were averaged for the final DJ coding. Ten cases were randomly selected to calibrate the coding. Inter-coder agreement on assigning codes to the DJ categories was 84.8%, and the process concluded with a discussion about differences and adjustments to the coding were made.

**Stable Agreements.** Stable agreement refers to the implementation of the peace agreement during the first two years after it was signed. Previous studies of a 16-case sample found that this variable plays a mediating role between PJ and the reconciliation component of DP as well as between equality and DP [2,21,43]. Fifteen of the 16 cases in those studies occurred between 1990 and 1997, where warring parties in a civil war reached a peace agreement and international actors were expected to play a large role in the implementation of the agreement [38]. The SA variable was assessed according to two components: (1) whether large-scale violence is brought to an end while peace-keeping implementers are present; and (2) whether war is terminated on a self-enforcing basis so that implementers can leave without war rekindling within a two-year time frame [19].

To provide a comparison with the previous study, we develop SA assessments based on these two components: whether violence ended and peacekeeping implementation. The violence component is based on the UCDP data set[2] regarding whether the peace agreement remained in force for five years after the agreement was reached and whether the conflict was active at the five-year mark ("stability"), although we adjust it for two years after the peace agreement was signed. Stability coding was performed using a 3 step scale: peace agreement failure (1), partial success (2), and complete success (3) in implementation of the agreement. A second coder's assessments differed from the first coder's assessments for only two cases, for a 96% rate of agreement.

The 50 case study includes 25 cases that involved peacekeeping operations, which allows us to evaluate the H2 hypothesis. For the cases with PK operations (PKO), we coded the actual number of years that the PKO remained in the country after the peace agreement was signed ("years of peacekeeping").[3] We also develop an index for peacekeeping length. This variable was coded on a three-point scale: PKO remained over 10 years or left due to violence/conditions in which peacekeeping could not continue (failure); PKO left in 2–10 years (partial success); PKO left under 2 years (success, with two years selected based on Downs and Stedman's [19] durability assessment).

**Durable Peace.** We assessed DP by evaluating four components during the eight years immediately following the signing of an agreement: extent of reconciliation, security institutions, governance institutions, and economic stability. The theoretical and empirical bases for this concept come from Diehl and Druckman's [16, 17] peace operation success framework and from Paris' [36] examination of peace building efforts after intra-state conflict. Additional sources were used to define the components of this index.

The reconciliation component encompasses backward-looking activities that seek to redress past offenses. This component includes efforts to address past crimes and establish truth commissions, but Thoms et al. [40] find that there is insufficient empirical evidence in the transitional justice literature to show that trials and truth commissions have state-level effects. We therefore combine elements such as reception

---

[2] Forty-two of the cases are included in the UCDP data set. For the remaining eight cases, a research assistant reviewed case data and assessed this variable.

[3] Some PKOs were in the country before the peace agreement was adopted, in which case this measure begins counting years when the agreement was adopted, and not total number of years of the PKO. In a few cases, the PKOs continue to the present day. The calculation for these cases goes to October 2014, when the variable was developed. The measure incorporates multiple PKOs in some cases, where successive PKOs were approved at the conclusion of the previous PKO.

and reintegration, community reconciliation and provisions for refugees into our reconciliation component [27]. The growing field of transitional justice focuses on these and related strategies to address the legacy of human rights abuses, with the objective of contributing to the foundation of societal peace.

Institution-building provides forward-looking components to build a durable peace. Paris highlights that democratic politics and capitalist economics "depend on public institutions to uphold rules, to maintain order, to resolve disputes impartially, and to regulate behavior incompatible with the preservation of market democracy itself" [36: p. 205]. We organize such institution-building efforts into three components: security institutions, governance institutions and economic institutions. Security institutions incorporate the extent to which military and police reform take place, demobilization and disarmament proceed, violent crime decreases, and ceasefire efforts hold within the country [17, 41]. The governance institutions component encompasses power sharing arrangements, rule of law and legal reform, electoral reform and holding elections, and human rights protection [14]. The economic component consists of conditions and actions that lead to economic stability, including measures of income equality, changing living standards, support programs from lending institutions, and the pursuit of conditions and policies for sustainable economic growth [5, 35].

To construct an index for the DP variable, a research assistant reviewed data in the PAM and UCDP databases, among other sources, for accounts of each country's experience after the peace agreement was signed. The cases were assessed for whether each of the 16 DP elements was mostly or very successful (1) or had limited success or failure (0). These assessments were added to develop the final "DP counts" score, which ranged from 0 successful DP elements to 16 successful DP elements for each case. We also added these assessments for each of the four DP components to develop codes for reconciliation, governance institutions, military institutions and economic stability. A calibration procedure was used to assess the replicability of the scores, with a second coder making independent assessments based on the case data. A discussion between the first and second coders then aimed at resolving any differences, with each code reconciled by discussion. The first and second coder's assessments matched on 86% of the DP components.

**Conflict Environment.** Exogenous factors related to the conflict itself could influence all of the variables under review in this study. The conflict environment (CE) can be measured based on the extent to which the population suffered during a conflict, leading to instabilities in the social fabric (internal conflict environment), and the extent to which external influences are contributing to or alleviating instabilities within the country (external conflict environment).

The three variables for the external CE were based on Downs and Stedman's [19] work on a "difficulty" measure. They explore the relationship among difficulty of the conflict environment and implementation success, and find four difficulty variables to be significant at the .01 level: existence of a spoiler, presence of disposable resources, presence of a hostile neighboring state, and presence of major power interest. Given that two of these variables are correlated – spoiler and hostile neighbors – we develop an index using three of the four, dropping the existence of a spoiler. Each variable was assigned a 1 or 2, and the external CE index was calculated as the average of the three scores.

The internal CE variable focuses to a large extent on the situation of a war-torn country's population prior to negotiation. It is based on three components, all assessed at the point that negotiations on the peace agreement began: months of violence, average number of deaths per month, and number of internally displaced persons (IDPs).

To establish which internal variables to include, we performed a factor analysis of five relevant variables in the PAM dataset. Two factors accounted for most of the variation. The first consists of very strong loadings for the number of months of violence, average deaths per month, and percent deaths of the population, with a moderately strong loading for months of violence. The second factor shows strong loadings for number of refugees and IDPs. A high correlation between average number of deaths per month and percentage of deaths of the population suggests that these variables would be interchangeable components for an internal CE index. Average deaths produces the weakest correlation with months of violence (.30 vs. .44 and .42 for number and % deaths), which indicates its usefulness for the CE index. Interestingly, we find that the number of refugees and number of IDPs are only weakly correlated (.34), and refugees correlates more strongly with months of violence than does IDPs (.71 vs. .47). Thus, IDP is a better indicator to include in the CE index. The three components that are selected for the CE index – months of violence, average number of deaths per month, and IDPs – are only weakly related to each other and thus not redundant.[4] An index was developed using a four point scale, in which the cases were assessed according to which quartile they represented for each of the three variables. More conflictual environments (highest quartile for months of violence, average deaths per year and number of IDPs) received the highest scores, while the least conflictual environments received the lowest scores. The points for each of the three variables were averaged to create a CE index.

## 6    Results

The justice and implementation variables assessed in our sampled cases are sequenced in time in the direction of PJ → DJ → SA → DP. Thus, a causal assumption is plausible and provides a basis for regression-based analyses supplemented with bivariate and partial correlations. The results are presented in the order of the hypotheses.

*H1: Stable agreements (SA) correlate with durable peace (DP). If a peace agreement has high (low) stability, then it will also have high (low) durable peace.*

Stable agreements strongly predict DP (beta = .55; t = 4.54, p < .0001). The bivariate correlation is identical to the beta at .55 (p < .0001). When the 50 cases are organized by whether their SA and DP scores are above or below the mean (see Table 1), the strength of the relationship holds across a variety of controls: for example,

---

[4] We evaluate the CE index based on absolute values and divided by population, but do not find different results with the population-adjusted figures. The correlations reported in the paper are based on the absolute values.

the SA-DP correlation controlling for the combined PJ/DJ variable is .54; when controlling for PJ, the correlation is .56; when controlling for equality, the correlation is .54. Further, factor analysis results show that both variables load strongly on the same factor. Thus, H1 is supported.

The next hypothesis considers peacekeeping missions as part of the implementation of the agreements.

*H2: Violence reduction is correlated with peacekeeping mission duration. If violence is reduced (increased), then peacekeeper missions will be of shorter (longer) duration.*

The correlation between the two parts of the Downs and Stedman [19] implementation index is .22 (p > .05). Further, length of PK missions do not correlate significantly with DP (r = .23). Thus, the two parts of the implementation index are relatively independent. As shown by the results reported for H1, violence reduction predicts, and is strongly correlated with, DP. Thus, DP is influenced more by the violence reduction than by the peacekeeping aspect of implementation.

*H3: Reconciliation and institutional change are correlated. If reconciliation is achieved (not achieved), then security and governing institutions will be strengthened (not strengthened).*

Strong correlations between reconciliation and each of the institutional components of DP were obtained: reconciliation and security (r = .61, p < .001) and between reconciliation and governing (r = .57, p < .001). Security institutions also correlated with the governing institutions component of DP (r − .53, p < .001). Factor analysis results show that the three components have strong loadings on the same factor. These results suggest that this is an inter-dependent cluster of DP components: each bi-variate correlation between two components is reduced when controlling for the third component (e.g., for the reconciliation-security correlation, from .61 to .45 when controlling for governing institutions). Thus, supporting this hypothesis, progress made on durable peace following agreements is indicated by progress on the composite of reconciliation, security and governing institutions. The question of what drives this progress is addressed in the hypotheses below.

*H4a: Economic growth does not occur during the short-term post-war recovery period.*
*H4b: The economic component of DP is uncorrelated with reconciliation and the development of security and governing institutions.*

Economic growth following the peace agreements is flat with an average score across the 50 cases of .64 on a scale that ranges from 0–3. The most change occurs on support from lending institutions (change occurs on 25 of the 50 cases) with either no change (improved living standards) or marginal change (policies for sustaining economic growth) on the other components. The correlations between economic growth and each of the other parts of DP are weak: economic growth and reconciliation (r = 0); economic growth and security institutions (r = .34); economic growth and governing institutions (r = .02). The correlation between economic growth and the DP index (excluding the economic component) is .15. Nor does economic growth correlate with the justice variables: PJ (.20); DJ (.15); equality (.15); combined PJ/DJ (.23). Factor analysis results show that the economic growth variable is isolated from the other variables used in the analysis: it loads strongly only on a factor that accounts for about 10% of the total variance explained in the correlation matrix. The other DP components have strong loadings on a factor that explains 50% of the total variance. Thus, supporting these hypotheses, economic growth did not occur during the recovery period and is not correlated with the other components of DP.[5]

*H5: The more central is the equality principle in the agreement, the more stable is that agreement.*

Equality predicts DP with a significant beta (.37; t = 2.73, P < .01). The moderate correlation between these variables (.37) provides modest support for the hypothesis. The correlations are weaker for each of the DP components: reconciliation (.33); security (.33); governance (.24), and economics (.15). Consistent with these results, equality has a moderate loading on the factor that accounts for 50% of the explained variance. The DP variables have stronger loadings on this factor (.51 versus .94, .79, .86, and .71 for DP, reconciliation, security, and governance respectively). The mediating role played by the equality principle is explored further with the H8 and H9a results presented below

*H6: The more central is the proportional principle in the agreement, the stronger is the economic growth in the society.*

The proportional principle of DJ does not predict economic growth during the recovery period. A beta of .11 is not significant. Neither does proportionality load on the key factor accounting for 50% of the variation. Thus, proportional outcomes do not relate to economic growth following the civil wars. This hypothesis is not supported.

*H7: The more that parties adhere to PJ principles during the process, the stronger is the reconciliation between them during the post-agreement period.*

PJ predicts reconciliation with a beta of .49 (t = 3.51, p < .001). PJ and reconciliation also load on the same factor (with loadings of .70 and .79 respectively), which accounts for 50% of the explained variation. The correlation is stronger for the 25 cases that used PK missions (r = .65). Supporting the hypothesis, this strong relationship between PJ and reconciliation was obtained as well in the 16-cases study by Wagner and Druckman [43].[6] Mediated relationships involving these variables are explored further with H9b below.

*H8: When the equality principle is emphasized in the agreement provisions, adherence to PJ principles lead to more stable agreements.*

Mediation analyses showed that equality did not mediate the relationship between PJ and stable agreements. Correlations between equality and PJ (.30) and between

---

[5] Correlations were also computed with a three-component DP index that excludes economic growth. The pattern of correlations between this index, SA, and the justice variables is only marginally higher than those reported with the four-component index. The two indexes are very highly correlated (.98).

[6] Interestingly the 16 cases used in Wagner and Druckman [43] also used PK missions following the agreements. The correlations between PJ and reconciliation were almost identical for the two PK data sets: .66 and .65.

equality and stable agreements (.15) are weak. Nor did equality mediate the relationship between PJ and DP ($z = 1.97$, $p < .12$, one-tailed). The correlation between equality and DP is .37.

*H9a: An emphasis on the equality principle leads to durable peace when the agreements are stable.*

Mediation analyses showed that stable agreements did not mediate the relationship between equality and DP. Although the correlation between stable agreements and DP was strong at .55, the relationship between equality and stable agreement was weak (.15) as noted above. Thus, this hypothesis is not supported.

*H9b: Adherence to PJ principles during the process leads to reconciliation when the agreements are stable.*

Mediation analyses showed that SA did not mediate the relationship between PJ and the reconciliation component of DP. Although the correlation between PJ and reconciliation was relatively strong at .49, the relationship between PJ and stable agreements was very weak at .06. Thus, this hypothesis is not supported.

*H9c: Adherence to PJ principles leads to reconciliation when negotiators adhere to DJ principles and the agreements are stable.*

This hypothesis posits two mediators: DJ and SA. A bootstrapping procedure was used to ascertain whether DJ and SA mediated the effects of PJ on reconciliation. The procedure was run with 1000 re-samples and a 95% confidence interval (see Hayes [26]: model 6). First, we evaluated the direct effects of each mediating variable. The results showed that PJ ($p < .008$) and SA ($p < .01$) were strong predictors of reconciliation. The DJ-reconciliation relationship reached borderline significance at $p < .09$. Second, we evaluated the indirect effects of these variables. Two significant models, referred to as serial mediation, were found. One consisted of the following path: PJ → DJ → Reconciliation. Statistical significance is indicated by the confidence intervals for the path (lower, .0002, upper, .0846); 0 is not included in the interval, indicating statistical significance. Another included SA in the path: PJ → DJ → SA → Reconciliation. The confidence intervals for this indirect model did not include 0 (lower, .0004, upper, .0440), thus indicating statistical significance. This path provides support for the hypothesis.

A final hypothesis posits a path that leads to durable peace:

*H9d: Adherence to PJ principles leads to durable peace when negotiators adhere to DJ principles and the agreements are stable.*

Using the same bootstrapping procedure, we first evaluated the direct effects of each mediating variable. The results showed that SA ($p < .0001$) and PJ ($p < .001$) were strong predictors of DP. Second, we examined the indirect effects of these variables. The best model consisted of a path from PJ to DP through DJ and SA. Statistical significance is indicated by the confidence intervals for this path (lower, .0050, upper, .2168), where 0 is not included in the interval. The serial path takes the following form:

$$PJ \rightarrow DJ \rightarrow SA \rightarrow DP$$

The path shows a linear relationship between negotiation process (PJ) and outcomes (DJ), on the one hand, and short-term (SA) and long-term (DP) peace on the other. Stable agreements and durable peace develop from adherence to justice principles during the negotiations and in the negotiated outcome. Neither of two shorter paths (PJ → DJ → DP or PJ → SA → DP) were significant; 0 was included in the range from the lower to upper confidence interval.[7] These results provide strong support for hypothesis H9d.

# 7  Discussion

Many of the findings reported in this article contribute in important ways to our understanding of how agreements to terminate civil wars contribute to peace, offering our analyst in the opening paragraph insights for advice to the negotiating parties. The implications of these findings are discussed in this section.

## 7.1  Stable Agreements and Durable Peace

Stable agreements preceded and predicted durable peace. This finding supports the earlier result obtained in a 16-case analysis of SA and DP by Wagner and Druckman, [43]. But it also provides clarification for the earlier results. Both correlations, with a 16 and 50-case sample, are due primarily to violence reduction rather than peacekeeping duration. Why is violence reduction a necessary (but not sufficient) condition? Only three cases in the sample attained some degree of DP without a stable agreement. The 1996 Philippines and 2002 Colombia cases were just above the DP mean (3.86) at 4. The 1999 case of Sierra Leone was considerably above the mean with a DP score of 9. This anomaly illustrates a difference between the short term assessment of SA and the long term assessment of DP. The 1999 agreement broke down quickly but was renegotiated a couple of years later with an agreement that was stable. The eight year DP period captures both the period after the first (1999) and second (2001) Sierra Leone agreements.

Eleven cases achieved stable agreements that did not eventuate in high levels of DP (see Table 1). For these cases, other conditions were needed to make the transition. One condition is justice during the negotiation process and in the outcome. Eight of the 11 cases were below the mean for PJ; five of the 11 cases were below the DJ mean for this variable. Another condition is reconciliation following the agreement: These eleven cases had the lowest scores on the reconciliation variable. The findings suggest that, although stable agreements were achieved in these cases, relational issues

---

[7] The shorter paths (single mediating variables) are stronger when evaluated with Sobel's z statistic. DJ is shown to mediate the relationship between PJ and DP at a borderline level of significance ($z = 1.54$, $p < .06$, one-tailed) as well as between PJ and reconciliation ($z = 1.43$, $p < .08$, one-tailed). SA also mediates the relationship between DJ and DP ($z = 1.65$, $p < .05$, one-tailed).

**Table 1.** Above and below the mean for DP [3.86] and SA [1.98].

| | Durable Peace | |
|---|---|---|
| | Above | Below |
| Stable Agreement | | |
| Above | Angola 2002 | Congo 1999 |
| | Bangladesh 1997 | Senegal 2004 |
| | Bosnia-Herzegovina 1995 | Sudan 2005 |
| | Burundi 2000 | Timor-Leste 1999 |
| | Djibouti 2001 | Haiti 1993 |
| | El Salvador 1992 | Cote D'Ivoire 2003 |
| | Guatemala 1996 | DR Congo 2002 |
| | Indonesia 2005 | Somalia 2008 |
| | Liberia 2003 | Cyprus 1964 |
| | Macedonia 2001 | India 1975 |
| | Mali 1992 | India 1988 |
| | Mozambique 1992 | [22%] |
| | Niger 1995 | |
| | Papua New Guinea 2001 | |
| | South Africa 1993 | |
| | Tajikistan 2005 | |
| | United Kingdom 1998 | |
| | Yemen 1990 | |
| | Slovenia 1991 | |
| | Kosovo 1999 | |
| | Nicaragua 1988 | |
| | [42%] | |
| | | |
| Below | Philippines 1996 | Cambodia 1991 |
| | Sierra Leone 1999 | Guinea-Bisou 1998 |
| | Colombia 2002 | Rwanda 1993 |
| | [6%] | Liberia 1991 |
| | | Mozambique 1984 |
| | | Papua-New Guinea 1990 |
| | | Somalia 1997 |
| | | South Africa 1978 |
| | | Philippines 2001 |
| | | Angola 1989 |
| | | Colombia 1986 |
| | | Lebanon 1989 |
| | | Philippines 1987 |
| | | Sri Lanka 1957 |
| | | Uganda 1985 |
| | | [30%] |

remained as reflected in the relatively low PJ and reconciliation scores. PJ is also shown to predict reconciliation suggesting that satisfying PJ principles during the negotiation process is likely to influence the chances for reconciling the conflict or, more generally, attaining a durable peace.

The eleven off-diagonal cases in Table 1 reduce the correlation between SA and DP (from .80 in the earlier 16-case study to .55 in this 50-case analysis). This is due largely to the 11 cases where SA did not lead to DP. These cases provided an opportunity to understand the role played by PJ in DP. They also reduced the correlation between PJ and SA: These cases were characterized by relatively low PJ and high SA. As a result, SA did not mediate the relationship between PJ and DP or reconciliation. The key, however, is the direct relationship between PJ and reconciliation. PJ may set in motion relational dynamics that facilitate reconciliation. When PJ principles are not adhered to, reconciliation is less likely to eventuate, even when the agreements are stable (high SA).

The role played by DJ, on the other hand, is indirect. It was shown to mediate relationships between PJ and reconciliation. Thus, the PJ-reconciliation relationship goes through negotiated outcomes: the path is from PJ through DJ to reconciliation. Stable agreements did not mediate the relationship between PJ and DP or reconciliation, due at least in part to the 11 cases where stable agreements were not accompanied by adherence to PJ principles or reconciliation.

Stable agreements do however play a mediating role in the path from PJ to reconciliation and to DP. Both DJ and SA serve to mediate the relationship between PJ and DP. Thus, the role of SA in producing durable peace is both direct and indirect. As noted above, a moderately strong correlation between SA and DP is due to the off-diagonal cases where high SA did not lead to high DP. The serial mediation results provide further insight. SA may work better when adherence to both PJ and DJ principles precede it. More generally, the serial mediation analyses show a time-lagged relationship between the four key variables in our study. Justice considerations precede and influence the prospects for reconciliation and peace following the termination of civil wars. Improved relationships (reconciliation) and changes in security and governing institutions (the other components of DP) are dividends of adhering to justice principles during the negotiation of peace agreements and in the outcomes of those negotiation processes. This is the key finding of this study.

## 7.2 Peacekeeping Missions

The cases divide evenly between those that did and did not have peacekeeping missions in the country following the agreement. Additional analyses showed that the duration of the missions after the agreement was signed varies with the intensity of the internal (but not the external) conflict: Shorter missions occur in lower conflict environments ($r = -.52$, $p < .007$). Not surprisingly, peacekeepers are more challenged when they must deal with continuing violence and deaths as well as with a large number of IDPs. The mission is less influenced by the state of the external conflict, including the presence of hostile neighbors, disposable natural resources, and the number of warring parties. The former are variables that peacekeepers must deal with on the ground. The latter consist of variables over which they have little control. This distinction is important when considering criteria for evaluating the success of peace operations. Containing violence within the population would seem to be a more appropriate goal than resolving the larger conflict [18]. Their chances of attaining success are improved when the internal conflict is manageable. Peacekeepers may be a relatively small cog in the wheel of progress on resolving the external aspects of the conflict. They must coordinate their efforts with peacemakers and peacebuilders.

The findings also provide insight into the role played by justice in the peacekeeping cases. Shorter missions occur when PJ principles are adhered to during the negotiation ($r = .58$, $p < .01$). This relationship is not influenced by the difficulty of the internal conflict environment: the correlations are practically identical with and without controls for the conflict environment. Further, relationships between the justice variables and DP are stronger for the PK than for the non-PK cases. This finding suggests a connection between negotiation process, duration of PK missions and durable peace. Mission length is shortened when parties adhere to PJ principles, increasing the prospects for DP.

This path may turn on consent. The idea is that the relationship dividends gained from adhering to PJ principles encourages the consent needed for PK missions, which, if successful (short duration), lays the groundwork for durable long-term peace. Although we did not include consent among our variables in this study, earlier research suggests that this may be a plausible path. In their taxonomic analysis of peacekeeping missions, Diehl et al. [18] showed that consent correlated strongly with the role of the peacekeeper and the conflict management process used by them. High levels of consent by the host state occurred when peacekeepers were in the role of third parties and the process was more integrative than distributive. This cluster of variables suggests a cooperation dimension. The findings obtained in our study further suggest that cooperation is set in motion by PJ during the negotiation.

## 7.3 Economic Stability

The economic stability component of DP did not correlate with the other components, reconciliation, security or governance institutions. Nor does it load on the same factor as these components. These findings replicate the results obtained with the smaller sample of cases analyzed by Wagner and Druckman [43]. They also support O'Reilly's [35] findings obtained with a different sample of cases. The lack of relationship between economic growth and peace, shown in the three studies, calls into question an economic peace dividend. This may be due, at least in part, to the time needed for the development of institutions that would facilitate the investments needed for economic recovery following civil wars as suggested by O'Reilly [35]. Eight years may be insufficient. If so, then a longer time period for observing economic development is warranted.

Although economic issues are relatively de-emphasized in the peace agreement provisions, the relative percentages of different types of issues in the provisions did not correlate with either the peace or justice variables. Nor does the conflict environment influence economic development: variation in conflict intensity was unrelated to variations in economic development over an eight-year post-agreement period [see 43].

This is not meant to imply, however, that the conflict environment is irrelevant. Rather, it is likely to influence economic growth in conjunction with institutional development. A relatively tranquil environment over a long period of time may provide a vital condition for the needed private investment that would stimulate the economy. Thus, the reemergence of hostilities between the former combatants would hamper progress toward economic stability and growth. This is a problem ripe for research, particularly with regard to understanding the interactive and recursive effects of these variables.

## 8  Conclusion

To answer our question of why so many peace agreements fail to achieve their goal of bringing peace to war-torn nations, we would reply that not enough attention is given to justice principles in the peace agreement negotiation and outcome. The findings of this research show that justice matters. PJ sets in motion a process that has both short and long-term impacts. These impacts are clear from the serial mediation results: PJ leads to just outcomes that produce stable agreements and durable peace. This path in the 50-case study is similar to the sequence found in the earlier 16-case analyses, bolstering the argument for generality. Both studies also show that the same progress made on reconciliation and institution building was not evident in economic growth. This is likely due to the time it takes private investors to acquire confidence in the stability of the economy following the civil war.

Justice matters also with respect to peacekeeping missions. Adherence to principles of PJ for the sample of PK cases led to shorter duration missions followed by durable peace. Conceivably the cooperation encouraged by PJ led negotiators to provide consent for the missions, which, in turn, increased the chances for both efficient and effective missions.

**Acknowledgments.**  Support for this study was provided by a Swedish Research Council grant [C0114201; registration number 421-2012-1142] for a research project titled, "From peace negotiation to durable peace: The multiple roles of justice." Special thanks go to Sayra van den Berg and Annerose Nisser for their excellent work on coding the documents for each case.

## References

1. Adams, J.S.: Inequity in social exchange. In: Berkowitz, L. (ed.) Advances in Experimental Social Psychology, vol. 2, pp. 267–299. Academic Press, New York (1965)
2. Albin, C., Druckman, D.: Equality matters: negotiating an end to civil wars. J. Conflict Resolut. 56(2), 155–182 (2012)
3. Albin, C., Druckman, D.: Procedures matter: justice and effectiveness in international trade negotiations. Eur. J. Int. Relat. 20(4), 1014–1042 (2014)
4. Ambrose, M.L., Cropanzano, R.: A longitudinal analysis of organizational fairness: an examination of reactions to tenure and promotion decisions. J. Appl. Psychol. 88(2), 266–275 (2003)

5. Ball, N., Halevy, T.: Making Peace Work: The Role of the International Development Community. Johns Hopkins University Press, Baltimore (1996)
6. Binningsbø, H.M., Loyle, C.E., Gates, S., Elster, J.: Armed conflict and post-conflict justice, 1946–2006: a dataset. J Peace Res. **49**(5), 731–740 (2012)
7. Brockner, J., Tyler, T.R., Cooper-Schneider, R.: The influence of prior commitment to an institution on reactions to perceived unfairness: the higher they are, the harder they fall. Adm. Sci. Q. **37**, 241–261 (1992)
8. Brockner, J., Wiesenfeld, B.M.: An integrative framework for explaining reactions to decisions: interactive effects of outcomes and procedures. Psychol. Bull. **120**, 189–208 (1996)
9. Brockner, J.: Scope of justice in the workplace: the case of survivors' reactions to co-worker layoffs. J. Soc. Issues **46**, 95–106 (1990)
10. Brockner, J., DeWitt, R., Grover, S., Reed, T.: When it is especially important to explain why: factors affecting the relationship between managers' explanations of a layoff and survivors' reactions to the layoff. J. Exp. Soc. Psychol. **26**, 389–407 (1990)
11. Burton, J.W.: The history of international conflict resolution. In: Azar, E.E., Burton, J.W. (eds.) International Conflict Resolution: Theory and Practice, pp. 40–55. Lynne Reinner, Boulder (1986)
12. Collier, P.: On the economic consequences of civil war. Oxford Econ. Pap. **51**(1), 168–183 (1999)
13. Cook, K.S., Hegtvedt, K.A.: Distributive justice, equity, and equality. Annu. Rev. Sociol. **9**, 217–241 (1983)
14. Cousens, E.M., Kumar, C., Wermester, K. (eds.): Peacebuilding as Politics: Cultivating Peace in Fragile Societies. Lynne Rienner, Boulder (2001)
15. Deutsch, M.: Distributive Justice: A Social-Psychological Perspective. Yale University Press, New Haven (1985)
16. Diehl, P.F., Druckman, D.: Peace operation success: the evaluation framework. In: Druckman, D., Diehl, P.F. (eds.) Peace Operation Success: A Comparative Analysis. Martinus Nijhoff Publishers, Leiden (2013)
17. Diehl, P.F., Druckman, D.: Evaluating Peace Operations. Lynne Rienner, Boulder (2010)
18. Diehl, P.F., Druckman, D., Wall, J.: International peacekeeping and conflict resolution: a taxonomic analysis with implications. J. Conflict Resolut. **42**(1), 33–55 (1998)
19. Downs, G., Stedman, S.J.: Evaluation issues in peace implementation. In: Stedman, S.J., Rothchild, D., Cousens, E.M. (eds.) Ending Civil Wars: The Implementation of Peace Agreements. Lynne Rienner, Boulder (2002)
20. Druckman, D.: Settlements and resolutions: consequences of negotiation processes in the laboratory and in the field. Int. Negot. **7**(3), 313–338 (2002)
21. Druckman, D., Albin, C.: Distributive justice and the durability of peace agreements. Rev. Int. Stud. **37**(3), 1137–1168 (2011)
22. Druckman, D., Wagner, L.: Justice and negotiation. Annu. Rev. Psychol. **67**, 387–413 (2016)
23. Harmon, D., Kim, P.: Trust repair via distributive justice rationales: the contingent implications of equity, equality, and need. In: 26th Meeting of the International Association of Conflict Management, Takoma, WA (2013)
24. Hartzell, C., Hoddie, M.: Institutionalizing peace: power sharing and post-civil war conflict management. Am. J. Polit. Sci. **47**(2), 318–332 (2003)
25. Hauenstein, N.M.A., Mcgonigle, T., Flinder, S.W.: A meta-analysis of the relationship between procedural justice and distributive justice: implications for justice research. Empl. Responsib. Rights J. **13**(1), 39–56 (2001)

26. Hayes, A.F.: Introduction to Mediation, Moderation, and Conditional Process Analysis: A Regression-based Approach. The Guilford Press, New York (2013)
27. Hayner, P.B.: Unspeakable Truths: Transitional Justice and the Challenge of Truth Commissions. Routledge, New York (2011)
28. Hollander-Blumoff, R., Tyler, T.R.: Procedural justice in negotiation: procedural fairness, outcome acceptance, and integrative potential. Law Soc. Inq. **33**, 473–500 (2008)
29. Kabanoff, B.: Equity, equality, power, and conflict. Acad. Manag. Rev. **16**, 416–441 (1991)
30. Kapstein, E.B.: Fairness considerations in world politics: lessons from international trade negotiations. Polit. Sci. Q. **123**, 229–245 (2008)
31. Lind, E.A., Tyler, T.R.: The Social Psychology of Procedural Justice. Plenum, New York (1988)
32. Markusen, A., DiGiovanna, S., Leary, M.C. (eds.): From Defense to Development? International Perspectives on Realizing the Peace Dividend. Routledge, London (2003)
33. Martin, P.: Coming together: power-sharing and the durability of negotiated peace settlements. Civ. Wars **15**(3), 332–358 (2013)
34. Mitchell, S.M., Diehl, P.F., Morrow, J.D.: Guide to the Scientific Study of International Processes. Wiley, New York (2012)
35. O'Reilly, C.: Investment and institutions in post-civil war recovery. Comp. Econ. Stud. **56**, 1–24 (2014)
36. Paris, R.: At War's End: Building Peace After Civil Conflict. Cambridge University Press, Cambridge (2004)
37. Shapiro, D.L.: The effect of explanations on negative reactions to deceit. Adm. Sci. Q. **36**, 614–630 (1991)
38. Stedman, S.J.: Introduction. In: Stedman, S.J., Rothchild, D.E., Cousens, E.M. (eds.) Ending Civil Wars: The Implementation of Peace Agreements. Lynne Rienner, Boulder (2002)
39. Szablewska, N., Bachmann, S.-D. (eds.): Current Issues in Transitional Justice: Towards a More Holistic Approach. Springer, Heidelberg (2015)
40. Thoms, O., Ron, N.T., Paris, R.: State-level effects of transitional justice: what do we know? Int. J. Transit. Justice **4**, 329–354 (2010)
41. United Nations: Report of the Panel on United Nations Peace Operations. United Nations (2000)
42. Wagner, L., Druckman, D.: The role of justice in historical negotiations. Negot. Confl. Manag. Res. **5**(1), 49–71 (2012)
43. Wagner, L., Druckman, D.: Drivers of durable peace: The role of justice in negotiating civil war termination. Grp. Decis. Negot. **26**(1), 45–67 (2017)
44. Zartman, I.W., Druckman, D., Jensen, L., Pruitt, D.G., Young, H.P.: Negotiation as a search for justice. Int. Negot. **1**(1), 79–98 (1996)

# Erratum to: Negotiating Peace: The Role of Procedural and Distributive Justice in Achieving Durable Peace

Daniel Druckman[1,2,3](✉) ⓘ and Lynn Wagner[4]

[1] George Mason University, Fairfax, VA, USA
dandruckman@yahoo.com
[2] Macquarie University, Sydney, Australia
[3] University of Queensland, Brisbane, Australia
[4] International Institute for Sustainable Development, Winnipeg, Canada
lynnwagner@jhu.edu

**Erratum to:**
**Chapter "Negotiating Peace: The Role of Procedural**
**and Distributive Justice in Achieving Durable Peace"**
**in: D. Bajwa et al. (Eds.):**
**Group Decision and Negotiation, LNBIP,**
**DOI: 10.1007/978-3-319-52624-9_12**

The version of this paper that was initially uploaded to SpringerLink was a preliminary version. This has now been corrected.

The updated original online version for this chapter can be found at
DOI: 10.1007/978-3-319-52624-9_12

© Springer International Publishing AG 2017
D. Bajwa et al. (Eds.): GDN 2016, LNBIP 274, p. E1, 2017.
DOI: 10.1007/978-3-319-52624-9_13

# Author Index

Printed in the United States
By Bookmasters

Printed in the United States
By Bookmasters